Budapest Scientific

Budapest Scientific

A Guidebook

István Hargittai
Magdolna Hargittai

OXFORD
UNIVERSITY PRESS

OXFORD
UNIVERSITY PRESS

Great Clarendon Street, Oxford, OX2 6DP,
United Kingdom

Oxford University Press is a department of the University of Oxford.
It furthers the University's objective of excellence in research, scholarship,
and education by publishing worldwide. Oxford is a registered trade mark of
Oxford University Press in the UK and in certain other countries

First Edition published in 2015
Impression: 1

Published in the United States of America by Oxford University Press
198 Madison Avenue, New York, NY 10016, United States of America

British Library Cataloguing in Publication Data
Data available

Library of Congress Control Number: 2014942678

ISBN 978–0–19–871907–6

Printed and bound by
CPI Group (UK) Ltd, Croydon, CR0 4YY

PREFACE

This book introduces the reader to the *visible* reminders of science and scientists in Budapest. These may be buildings, statues, plaques, or there may be some other reason to bring up a related topic. The English word "science" means what other languages express as "natural" sciences. We will be using the term science primarily in this sense, but will include visible reminders in technology, inventions, medicine, humanities, and social sciences as well.

It is hard to say whether Budapest has a special aura for science. It may well have. Albert Szent-Györgyi, when he visited Budapest one more time shortly before he died, declared that he came because he wanted to breathe, once again, the atmosphere of respect for science. Budapest has been the venue for numerous scientific achievements and the cradle of many individuals who became great names in science in Hungary and, often, far beyond its borders.

The visible reminders of science and scientists cover a time scale from the turn of the eighteenth and nineteenth centuries to the present day, with only a few exceptions from earlier periods. The reform era in the first half of the nineteenth century was very progressive. The period between 1867 and 1914 in Hungary was an especially fertile time for science and for many other fields of human endeavor, including public education. World War I, the revolutions, and the ensuing White Terror and Admiral Horthy's autocratic and anti-Semitic reign were not a time conducive to great science. A significant number of scientists left the country, many of whom then became successful in Western Europe and the United States of America. The *numerus clausus* law of 1920 and the brutal anti-Jewish legislation of the late 1930s and early 1940s culminated in the Hungarian Holocaust which was also a tragedy for the Hungarian scientific community.

After World War II and a couple of years of democracy, forty years of communist dictatorship and Soviet domination followed. The repression was not uniform and the hardest early period later gave way to a milder regime. This softening was in part a consequence of the anti-Soviet revolution in 1956, because memories of this reminded the authorities that there were limits to the tolerance of those upon whom they imposed their rule. The gradual easing of travel restrictions was a typical feature of the hesitantly relaxing communist regime; from virtually no travel in the early 1950s to almost free travel by the time communism collapsed between 1989 and 1990. Scientists, along with musicians and sportsmen, had better opportunities to travel, even in the hardest of times, than did the rest of the population. With almost no other outlet for individual incentives, scientific research was a unique area of human endeavor. Under the democratic conditions of the next two decades, following the political changes of 1989–1990, science had to compete with all other areas for the best young people. We have yet to see the impact on science of recent developments in autocratic governance and the attempted restoration of the political regime of the period between the two world wars. The exodus of a mobile young population is an ominous sign.

Science administrators have praised the achievements of Hungarian scientists. This is an appreciation of the efforts of individuals in spite of much worse conditions for research than in developed nations. On the other hand, politicians may view this as evidence that, even at a low level of support, science in Hungary can thrive. A special feature in the evaluation of Hungarian science is the handling of the contribution of scientists who left their homeland and made their discoveries abroad. By incorporating their performance into the total, the evaluation of performance versus the extent of support is misleading. This caveat is needed, because our narrative, too, includes émigré scientists along with those that stayed in Hungary.

Another caveat concerns the character of coverage. Its primary base is the visible—physical—memorabilia. Their existence is not at all exclusively related to the greatness of achievements. Sometimes, because of an enthusiastic pupil or other devotee, someone gets a statue or a plaque while a comparable peer has no such remembrance. Thus, it is an important caveat that science history should not be compiled solely on the basis of such memorabilia.

———◆———

Our book has a simple structure. Each of its eight chapters carries the name of an individual, to symbolize the contents and aspirations of the chapter. The respective subtitles assist in delineation of the contents of the chapters, but those contents are defined loosely, hence, overlaps are unavoidable.

The spelling of names follows the Hungarian usage even if first names occur in translation in international literature. The order of names, first name first and surname second, follows the international usage, except in the names of streets and squares, where the original order is observed.

We give exact street addresses for all statues, plaques, and buildings that we introduce. They are spread all over Budapest and it would be of most use when searching for them to have a detailed map of Budapest to hand. In case the map happens to be in Hungarian, we append a miniscule dictionary showing key words that may be needed to use such a map.

———◆———

All photographs are by the authors unless otherwise indicated. They are all protected by copyright. For images from other sources, we have made reasonable efforts to uncover the copyright holders and contact them for permission to reproduce. In cases where this was not successful, we would appreciate any further information which would augment our efforts.

———◆———

Chapter Overviews

1 Albert Szent-Györgyi—Romanticism and Reality

The Nobel Prize has become a household name in discussions on Hungarian science. Albert Szent-Györgyi was one of two Nobel laureate scientists who did much of his prize-winning research in Hungary (the other being Georg von Békésy) and he was the only one who was living in Hungary at the time that he received the Nobel Prize.

The popular notion of the inflated number of Hungarian Nobel laureates is both a blessing and a curse. It is a blessing because it has added to the international reputation of Hungarian science, but it has a negative impact when we ourselves believe in such an inflated number. In reality, some of the Nobel laureates hardly had any roots in the country while others were forced to abandon it, and they achieved their seminal contributions elsewhere. An honest approach counters the popular belief in the inflated numbers, and helps us to appreciate the true achievements that have emerged within Hungary's borders.

2 István Széchenyi—Learned Society

István Széchenyi appears here in the context of the Hungarian Academy of Sciences, and symbolizes our effort with this book. In 1825, Széchenyi and a few other aristocrats committed considerable portions of their wealth to found a learned society. Széchenyi did far more than donating monies; he wrote books and intiated railways, steamboating on the Danube, and a banking system for the country. The great Hungarian statesman, Lajos Kossuth—with whom his opinion often diverged—called Széchenyi *the Greatest Hungarian*.

The headquarters of the Hungarian Academy of Sciences stands on Széchenyi István Square in the center of Budapest. It is the focal point of Hungarian science regardless of whether the respective research is being carried out at the institutes of the Academy, at universities, or in other research organizations.

3 Loránd Eötvös—Universitas

Loránd Eötvös was a great physicist, of international renown, whose achievements in classical physics paved the way for Albert Einstein's theory of general relativity. He was a distinguished educator, who followed in his father's footsteps: József Eötvös was a writer and politician with a special flair for educational policy. This chapter focuses on science education, including what Loránd Eötvös University is today. The early history of the predecessors of this school is the history of the foundation for higher education in Hungary.

4 Ignác Semmelweis—Medicine

Ignác Semmelweis was the "Savior of Mothers," and many commemorations honor him. His grave is in the yard of the Museum of Medical History in Buda though he has an impressive tombstone at the Fiume Avenue National Necropolis. The campuses, clinics, and other institutions of Semmelweis University abound with busts and plaques of famous professors and other contributors to the medical profession.

5 Pál Kitaibel—Agricultural Science

The polymath Pál Kitaibel was primarily a botanist. He discovered a new element, but did not insist on his priority when he learned about the pre-eminence of others in this discovery. This chapter is about the memorabilia in botany, zoology, and agricultural science that have strong traditions in Hungary. Their principal venues are the veterinary school, the school of gardening and the food industry, the ministry of agriculture, the Museum of Agriculture, the Budapest Zoo, and the Botanical Garden.

6 Theodore von Kármán—Engineering and Innovation

Theodore von Kármán—Tódor Kármán—was an engineer and aerodynamicist, and one of the celebrities of twentieth-century science. He completed his university education and started his research career in Budapest. He was famous for seminal inventions and discoveries and he provided the scientific background for the air force of the US Army during World War II and for the US Air Force, later on.

Many celebrated Hungarian engineers and inventors trained at the Budapest University of Technology and Economics and its predecessors. We include memorabilia of British inventors, an American pioneer of the personal computer, Rubic's Cube, foundrymen, even the inventor of the ball-point pen, and many others.

7 László Rátz—Schools of Creativity

László Rátz was a high school mathematics teacher who gave private lessons to John von Neumann and provided reading books to Eugene P. Wigner. During the "happy peace time," between 1867 and 1914, Budapest was one of the fastest developing cities in the world. The famous Hungarian high school—the gimnázium—developed, and the gimnáziums became the workshops of creativity.

It was not just one gimnázium, but several that produced future internationally renowned scientists. Seven future Nobel laureates graduated from five of these gimnáziums. Three of the gimnáziums served as alma maters for the five "Martians," the famous émigré scientists—Theodore von Kármán, Leo Szilard, Eugene P. Wigner, John von Neumann, and Edward Teller—who risked their scientific careers in their efforts to defend the Free World from Nazi Germany in World War II and from the Soviet Union during the Cold War.

8 Gedeon Richter—Martyrs and Saviors

Gedeon Richter was the founder of the Hungarian pharmaceutical industry and one of the countless numbers of victims of the Hungarian Holocaust. Imre

Bródy was a pioneer of incandescent lamp research and industry and was killed in the course of deportation. These are but two examples of the losses that we remember. There are moving memorials to the Hungarian Holocaust for the general population. One of them, the "Shoes on the Bank of the Danube" is within walking distance of Széchenyi Square and the Hungarian Academy of Sciences. There are also memorials to the Swede, Raoul Wallenberg, the Hungarian Gábor Sztehlo, and a few others who saved Jews, including future luminaries of science, from annihilation. Alas, there are no specific memorials of the scientist–martyrs of the Holocaust. This creates the false impression that the Holocaust had no effect on Hungarian academic life.

———— • ◆ • ————

This book grew out of our love of science and our love of Budapest. We are happy to share this highly personal account with our readers.

István and Magdolna Hargittai
Budapest, January 9, 2015

ACKNOWLEDGMENTS

We have received kind and efficient assistance of most diverse kinds from many individuals in the preparation of the manuscript for this book. Their names are listed here with their cities indicated, except for Budapest.

Gerald L. Alexanderson (Santa Clara, CA), Endre A. Balazs (Fort Lee, NJ), Krisztina Batalka, Mihály and Piroska Beck, Lóránt Bencze, Erzsébet Besze, Ádám and Ágnes Böhm, Aaron Ciechanover (Haifa), Endre Czeizel, Janet Denlinger (Fort Lee, NJ), Ninette Dombrowsky, József Farkas, Tamás Fenyvesi, Tibor Frank, István Gera, Éva Gergő, Sándor Görög, György Haraszti, Balazs Hargittai (Loretto, PA), Eszter Hargittai (Evanston, IL), Diána Hay, Miklós Hernádi, Ferenc Hudecz, István Isépy, Éva Jeremiás, István Karádi, Géza Komoróczy, László Kovács, Jr., László Kovács (Szombathely), Mária M. Kovács, László Kúnos, György Liptay, Klára Majoros, Benjamin Makovecz, Roger Malina (Richardson, TX), the late György Marx, Mrs. György Marx, János Megyeri, Miklós Mohai, László Molnár, Péter Tibor Nagy, József Németh, George A. Olah (Los Angeles, CA), David Peleg (Kibbutz Dalia, Israel), Szilvia Peremiczky, János Philip, György Pokol, John C. Polanyi (Toronto), Imre Romics, Zoltán Schay, Vera Silberer, András Simonovits, Zsolt Szabó, Andrew Szent-Györgyi (Waltham, MA), Ágoston Szél, János Szépvölgyi, Vera T. Sós, Zsolt Tulassay, Charles Upton (Hudson, OH), Izabella Vajda, László Vámhidy, József Varga, Marina von Neumann Whitman (Ann Arbor, MI), Richard Wiegandt, the late Martha Wigner Upton, László Wojnárovits, and Ahmed Zewail (Pasadena, CA). We owe special thanks to Bob Weintraub and Irwin Weintraub (Beersheva) for scrutinizing the entire text and for useful suggestions. We express our appreciation to Sonke Adlung, Senior Editor (OUP), for his encouragement, and Victoria Mortimer, Senior Production Editor (OUP), for her dedicated efforts in bringing this book to life.

CONTENTS

1

Albert Szent-Györgyi

ROMANTICISM AND REALITY

Albert Szent-Györgyi lecturing about his Nobel-Prize-winning discoveries.

Courtesy of the Hungarian National Museum.

Albert Szent-Györgyi is renowned for his achievements in the biomedical sciences and the discovery of Vitamin C. He was a realist in science, but he was also a romantic dedicated to a wide variety of causes. One of his close associates at the University of Szeged, Ferenc Guba, joined him while Guba was still a student. He remembered:[1]

> In the lab, he [Szent-Györgyi] was an absolute realist. He once told me, 'I enjoy muscle research, but my dream would be brain research. Do you know why I don't do it? The reason is that what we know about the brain today—and our technical capabilities—would not allow me to reach truly important results.' This was his realism. You may be gifted, and you may have dreams, but if you pursue

unrealistic goals, you may end up with nothing. Outside the lab, however, his political maneuvering in his attempts to help Hungary get out of the German alliance was completely amateurish. At one point he traveled to Turkey—ostensibly to give a lecture, but in reality to negotiate with the British—and ended up talking with German agents posing as British officers.

Guba then added, "Szent-Györgyi may have been a romantic, but he could not be seduced by the temptation of power or wealth. He counted himself to be a Hungarian to the end, but he never painted himself in the Hungarian tricolor of red, white, and green. What I remember most vividly from my time with him is the free spirit in the lab."[2]

Albert Szent-Györgyi's nephew and long time co-worker, Andrew Szent-Györgyi of Brandeis University, also spoke about his uncle's romanticism:[3]

> He always wanted to study interesting things, and he believed that what he was working on at any given moment was the most interesting problem in science. His attitude may be best characterized by his 'fish story.' Lajos Zilahy, a Hungarian writer who also immigrated to the United States in 1947, was Albert's close friend and a passionate angler. He induced Albert to become interested in fishing, but for years, Albert couldn't catch anything. He told me, 'Whenever I go fishing, I use a big hook so that the fish I don't catch should be a big one.'

There are memorials in Budapest honoring science and scientists in general. We mention here five. A modern statue "Thinker" stands in front of an office building at Medve Street, District II. This statue reminds us of what Albert Szent-Györgyi said about how discoveries happen, when one sees what everybody else does but thinks what nobody had thought before.[4] A symbolic figure, "The Scientist," decorates the tombstone of the grave of engineer Adolf Czakó (1860–1942) at the Fiume Avenue National Necropolis. Czakó was a professor of the Budapest Technical University and its one-time rector. The statue symbolizes a scientist and cannot be

"Thinker" (by Imre Varga, 1968) at 29 Medve Street, District II.

Adolf Czakó's representation whose face along with his son's face is depicted on the side of the tombstone. The old bespectacled man of the statue is reading a large book and the book is supported by an owl—the symbol of wisdom.

Another statue that symbolizes science is on the top of the former "Kuria" (appeals court) building, now the Museum of Ethnography on Kossuth Lajos Square. The building was designed by the architect Alajos Hauszmann. The Museum of Ethnography was founded in 1872 as a subdivision of the Hungarian National Museum, and moved into its present building in 1973. It has a large collection based on the early cultural history of Hungarian and also a broad range of international regions of the world.

"The Scientist" by Jenő Bory as part of the tombstone of the grave of Adolf Czakó and his family at the Fiume Avenue National Necropolis (plot 11-5-9).

Photo by and courtesy of József Varga, Budapest.

The fourth example is a relief, "Science," on the façade of the headquarters of the Hungarian National Bank. Architect Ignác Alpár (see Chapter 5) designed the bank building.

The Museum of Ethnography, 12 Kossuth Lajos Square, District V.

A statue on the top of the Museum of Ethnography symbolizes *Science*.

The architect Alajos Hauszmann's relief (by János Pásztor, 1913) is on the wall of the building.

Corner section of the headquarters of the Hungarian National Bank at 8–9 Szabadság Square, District V.

Heroes' Square, at the end of Andrássy Avenue, next to City Park, has several relevant statues. Andrássy Avenue and Heroes' Square together are part of UNESCO's World Heritage.

King "Bookish" Coloman (in Hungarian, "Könyves" Kálmán, who reigned from 1095 to 1116) was a learned and progressive monarch; literature and the legal

Heroes' Square, at the end of Andrássy Avenue, at the border of Districts VI and XIV.

The statue at the top of the right colonnade symbolizes Knowledge and Glory.

system flourished under his rule, hence his nickname. He went down in history as an enlightened ruler who did away with the witch trials as he declared, "there are no witches."[5] He is depicted as making this statement, on the relief under his statue on Heroes' Square. Coloman suffered from physical handicaps, was "half-blind, a hunchback, crippled and a stutterer,"[6] but this did not prevent him from making significant conquests for his country.

King Matthias (Matthias Corvinus, in Hungarian, Mátyás, reigned 1458–1490) was a national king in that he did not inherit the throne, but rather, he was elected. He was a Renaissance sovereign, a patron of the arts and science. His statue stands on Heroes' Square. The relief beneath his statue depicts a scene in which architects present their design to Matthias.

The mathematician, astronomer, and astrologer of European fame, Johannes Müller (1436–1476), originated from Bavaria. Regiomontanus is the Latin version

Statues of Coloman (*left*) and Matthias (*right*) on Heroes' Square.

Left relief: King Coloman declares that there are no witches. *Right relief*: King Matthias amongst his scientists.

of his name and after his death, he became known by this name. He translated Ptolemy's second-century treatise *Almagest* from the Greek original into Latin which Copernicus and Galileo used as a textbook. At King Matthias's invitation, Regiomontanus arrived in Hungary in 1467 to serve as Royal Librarian. In 1471, he returned to Bavaria and settled in Nurnberg where he established a printing house. This happened soon after Johannes Gutenberg invented movable type thereby giving great stimulus to book printing.

The Italian humanist Antonius Bonfinius (Antonio Bonfini, 1434–1503) served at one time as King Matthias's court historian. He authored the ten-volume treatise *Rerum Hungaricum Decades* about the history of the Hungarians.

Relief of Regiomontanus in his working environment (by József Rátonyi, 1979). The memorial is in the garden of the Historical Museum of Budapest, at 2 Szent György Square, District I, in Buda Castle.

Left: Bonfinius's bust in Buda Castle, District I. The original (by Zoltán Farkas) was destroyed in World War II. István Fáskerti created the copy (2008).

Below: Stone of Science on the sidewalk at 1 Váci Avenue, District VI. The names of noted Hungarian scientists are carved into this huge piece of stone; the names of Nobel laureates are in gold.

There is a huge stone commemoration for scientists at the very start of Váci Avenue which has names of many important Hungarian scientists carved into it. The carvings are visible only at short range. According to popular opinion, this monument has not enhanced the beauty of its surroundings; some have called it a behemoth.

Facts and fantasy—realism and romanticism—blend in the popular notion of the extraordinary number of Nobel laureates from Hungary. However, the number would be very small if only those were included who lived a considerable part of their lives in Hungary, received their education in Hungary, and conducted at least a portion of their research in Hungary. There is no all-purpose single number, but in any case, it is more realistic to speak about Hungarian Nobel laureates and Nobel laureates with Hungarian roots. Two bona fide criteria for counting someone as a Hungarian Nobel laureate are Hungarian citizenship at the time of the award and whether the laureate did his (no "her" yet) award-winning work in Hungary.

There are three memorial tablets near the entrance of the building of the Association of Technical and Scientific Societies (METESZ) at 68 Fő Street, District II.

Federation of Technical and Scientific Societies (METESZ) at 68 Fő Street, District II.

Two tablets display the names of "Our Nobel laureates," one for the twentieth century and the other for the twenty-first century. The list is chronological and below we follow its order. However, we have omitted Herta Müller and inserted D. Carleton Gajdusek. It is debatable whether or not the addition of Gajdusek is justified (see below). In contrast, the inclusion of Herta Müller, Nobel laureate in Literature for 2009, among Hungarian Nobel laureates cannot be justified. She was born in 1953 in western Romania, of German-speaking parents. She wrote in German, and she left Romania in 1987 for West Germany, as it was then. The

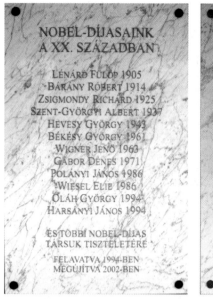

Memorial plaques dedicated to "Our Nobel Laureates" on the wall of the METESZ building. *On the left,* laureates of twentieth century are listed; *on the right,* those of the twenty first century.

region where she was born used to belong to the Kingdom of Hungary, but this is hardly a valid reason for claiming her among Hungarian Nobel laureates.

Even though the list is exaggerated, excluding any of the entries (beyond Müller) would not be appropriate. Instead, we briefly discuss the entries one at a time.

"Our Nobel Laureates"

Philipp von Lenard (1862–1947), Physics 1905, for his work on cathode rays. He was born in Pozsony, Hungary, known as Pressburg in German, which is today Bratislava, the capital of Slovakia. His family originated from the Austrian Tyrol. He studied in Vienna, Budapest, and Heidelberg. He worked for half a year in Budapest, and spent the rest of his career in Germany. In 1897, he was elected to membership of the Hungarian Academy of Sciences as a Hungarian citizen; he later became a German citizen. Von Lenard's political views have no relevance to whether or not he belongs to this list, but it is noteworthy that he was a vocal member of the Nazi Party and was infamous for his virulent anti-Semitism and opposition to Einstein's and others' "Jewish physics."

Philipp von Lenard on a Swedish stamp (on the left; Adolf von Baeyer, the 1905 chemistry Nobel laureate is in the front).

Robert Bárány.
Courtesy of Anders Bárány, Skane-Tranas, Sweden.

Richard A. Zsigmondy on Austrian and Hungarian stamps.

Robert Bárány (1876–1936), Physiology or Medicine 1914, for his discoveries concerning the vestibular apparatus. He was born in Vienna and never lived in Hungary. His father had moved from Hungary to Vienna. All non-Hungarian sources list him as Austrian. He spelled his name in the Hungarian way with the two "á" letters. His grandson, the Swedish Anders Bárány, professor emeritus of physics, Stockholm University, and former long-time secretary of the Nobel Committee of Physics, has kept the Hungarian way of writing his surname.[7]

Richard A. Zsigmondy (1865–1929), Chemistry 1925, for his pioneering work in colloidal chemistry. He was born in Vienna and never lived in Hungary. His surname is Hungarian; his parents had moved from Hungary to Vienna. The Austrian and Hungarian post offices have honored him with similar stamps.

Albert Szent-Györgyi von Nagyrapolt (1893–1986), Physiology or Medicine 1937, for his discovery of Vitamin C and other contributions. He was born and studied in Budapest. He conducted a considerable proportion of his award-winning research in Hungary, at Szeged University, in addition to working in British and Dutch laboratories. He was a Hungarian citizen when he received the Nobel Prize.

George de Hevesy (1885–1966), Chemistry 1943, for the introduction of isotopes as tracers in the study of chemical processes. He was born in Budapest. He studied in Budapest and at German universities. In 1920, he left Hungary and did his research, including the prize-winning research, in England, Germany, and Denmark. At the time of his Nobel award, he held Hungarian citizenship.

Georg von Békésy (1899–1972), Physiology or Medicine 1961, for uncovering how the cochlea works. The cochlea is the auditory portion of the inner ear. Békésy was born in Budapest, but spent most of his childhood and youth abroad. He earned his doctorate in physics at Budapest University. From 1924, he was engaged in research at the experimental station of the Royal Hungarian Postal Service. In 1946, he moved to Sweden, then, soon, to the United States. He received his Nobel distinction as a US citizen, but the research he performed in Hungary constituted a substantial part of his award-winning discoveries.

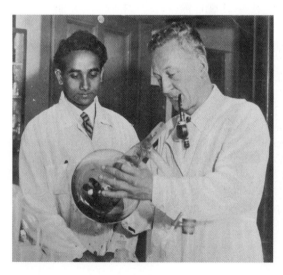

Albert Szent-Györgyi in the mid-1930s at Szeged University with a co-worker from India.

Courtesy of Andrew Szent-Györgyi, Waltham, MA.

George de Hevesy.
Courtesy of the late George Marx.

Georg von Békésy.
Courtesy of the late George Marx.

Eugene P. Wigner (1902–1995), Physics 1963, for both his discoveries in and application of the symmetry concept to nuclear physics. Wigner was born in Budapest and studied there through his college freshman year. In 1921, he left for Berlin and returned to Budapest for just one year. He did his research, first in Germany and then in the United States. In 1937, he became a US citizen.

Dennis Gabor (1900–1979), Physics 1971, for inventing holography. Gabor was born and studied in Budapest, but he did much of his university study in Berlin. In 1934, he moved to the United Kingdom, and he discovered holography in England. He received the Nobel Prize as a British subject.

D. Carleton Gajdusek (1923–2008), Physiology or Medicine 1976 (jointly with Baruch S. Blumberg), for discovering and understanding new kinds of infectious diseases. Gajdusek was born in 1923 in Yonkers, NY, from parents

Eugene P. Wigner.
Courtesy of the late Martha Wigner Upton.

Dennis Gabor.
Courtesy of the late George Marx.

Carleton Gajdusek in 1999 at the Budapest Technical University.

of Slovakian and Hungarian origin. He himself referred to these roots, but they had no impact on his career.

John C. Polanyi (1929–), Chemistry 1986 (jointly with Dudley R. Herschbach and Yuan T. Lee), for understanding the mechanism of elementary chemical reactions. Polanyi was born in 1929 in Berlin and in 1933 moved to England with his parents, Michael Polanyi and Magda Polanyi. He stresses that his parents' Hungarian roots did not have any impact on his scientific life. Polanyi studied in England and has had his scientific career in Canada.

Elie Wiesel (1928–), Peace 1986. He was born in 1928 in Sighetul Marmatiei in Romania, which before the 1920 Trianon Peace Treaty had been part of the Kingdom of Hungary (Máramarossziget). He attended secondary school in Debrecen, Hungary. By the time of World War II, the region where his family lived, for a few years became part of Hungary again; the Hungarian and German authorities deported Wiesel and his family to Auschwitz. There, his parents and younger sister perished. Wiesel survived the Holocaust, studied in Paris, and eventually moved to the United States. Most sources list him as a Romanian–French Nobel laureate. In 2004, Wiesel received a high Hungarian state award, which he renounced in 2012 when Hungarian state officials honored the memory of a pro-Nazi writer.

John C. Polanyi in 2006 in Budapest, in the Hargittais' home.

George A. Olah (1927–), Chemistry 1994, for initiating a new area in chemistry involving carbocations. Olah was born and completed all of his education in Budapest. He started his research career in Budapest, and left following the suppression of the Revolution in 1956. First he moved to Canada, then to the United States. He received his Nobel Prize as an American citizen, but has often expressed his pride in his Hungarian roots.

John Harsanyi (1920–2000), Economic Sciences in Memory of Alfred Nobel 1994 (jointly with John F. Nash and Reinhard Selten), for their work on noncooperative games. Harsanyi was born in Budapest and did all his schooling in Hungary. He graduated as a pharmacist, but received his doctorate in

Marion Wiesel and Elie Wiesel in 2004 at the Nobel Prize ceremonies in Stockholm.
Courtesy of Aaron Ciechanover, Haifa, Israel.

philosophy in 1950. He left Hungary, and studied economics in Australia. He embarked on research of game theory in Australia and continued this in the United States. He received his Nobel award as a dual Australian and US citizen.

George A. Olah in 1995 at the Budapest Technical University.

John Harsanyi.
Courtesy of the late George Marx.

Imre Kertész (1929–), Literature 2002, for his book *Fateless.* Kertész was born in Budapest. He survived the Holocaust and became a writer. He received the Nobel Prize as a Hungarian writer living in Hungary.[8]

Imre Kertész (second from right) in the company of György Ligeti, Hungarian–Austrian composer and musicologist (on the far left) and his wife, Vera Ligeti (on the far right), and the Hungarian-Swedish tumor biologists Eva and Georg Klein.
Courtesy of Eva Klein.

Avram Hershko in August 2004 in Woods Hole, MA, two months before the announcement of his Nobel Prize.

Avram Hershko (1937–), Chemistry 2004 (jointly with Aaron Ciechanover and Irwin Rose), for their discovery of protein degradation mediated by ubiquitin. Hershko was born in Karcag, Hungary. In 1944, he was deported together with his mother and brother, but survived the Holocaust. After the war, he and his family lived in Budapest. In 1950, they immigrated to Israel, where he completed his education. He has conducted his research mostly in Israel and also in the United States.

The above enumeration indicates the difficulty of determining the number of Hungarian Nobel laureates. Sadly, Hungary had chased away many on this list and embraced them only after they had received the highest accolades for work performed outside of Hungary.

The METESZ has another marble plaque, displaying a list of names of other significant Hungarian scientists, "Our Greats in Science." Similar plaques to those of the METESZ headquarters are on the new campus of Eötvös University at 1/A Pázmány Péter Walkway, District XI (see Chapter 3). The list of "other" scientists contains the following names: János Bolyai, Zoltán Bay, Loránd Eötvös, Ányos Jedlik, Theodore von Kármán, Sándor Körösi Csoma, John von Neumann, Ignác Semmelweis, János Szentágothai, István Széchenyi, Leo Szilard, and Edward Teller. They are not all equally well known, but we will meet most of these names in subsequent chapters. The list of "other scientists" could be criticized only for omission, not for inclusion.

———— • ◆ • ————

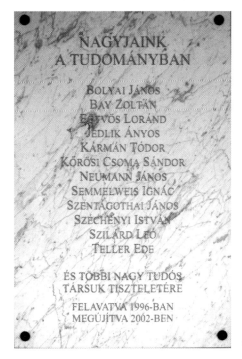

Memorial plaque dedicated to "Our Greats in Science" on the wall of the METESZ building.

The emphasis on the Nobel Prize is so conspicuous that some additional caveats would seem to be justified. First, there is no Nobel Prize in all fields of science, and even the Royal Swedish Academy of Sciences has recognized the need to diminish the gap caused by the absence of a Nobel Prize in some fields. There is, for example, the Crafoord Prize of the Royal Swedish Academy of Sciences, an award recognizing exceptional achievements in mathematics, geosciences, biosciences, and astronomy.

In 2003, the Norwegian Academy of Science and Letters established the Abel Prize in Mathematics, whose monetary reward is comparable with that of the Nobel Prize and which has high prestige in the academic community. Two mathematicians with Hungarian roots have already received this award: in 2005, the American citizen Peter Lax (1926–) and in 2011, the dual Hungarian–American citizen Endre Szemerédi (1940–).

Abel Prize-winner mathematicians Endre Szemerédi and Peter Lax in 2008 in Fort Lee, NJ.

The writer and journalist Arthur Koestler's (1905–1983) writings reveal a great deal of science. In his book, *The Act of Creation*, he provides a refreshing discussion on the nature of scientific discovery.[9] He describes the relationship between laughter and discovery and the fruitfulness of bringing different planes of thought together in finding something new. It is more than thought association; he calls it bisociation.

Beyond what Koestler wrote about science and his analysis of scientific discovery, he engaged in science popularization at the highest level. His book *The Sleepwalkers* is an account of the history of cosmology and astronomy in western science.[10] Its highlight is the section about Johannes Kepler, which appeared also as a separate book, *The Watershed*.[11] Koestler achieved worldwide fame with his anti-Communist novel, *Darkness at Noon*.[12] It is one of the best political novels of the twentieth century, and has influenced scientists, such as, for example, Edward Teller, and the science writer Martin Gardner.

Above: Arthur Koestler.
Courtesy of the Petőfi Literary Museum, Budapest.

Left: The statue of Arthur Koestler by Imre Varga (2009) on Lövölde Square, District VI, close to his childhood home.

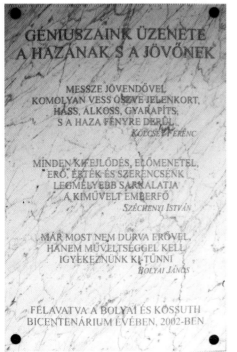

Plaque at the entrance to METESZ with messages by Hungarian geniuses to the Homeland and the Future.

19

For a moment, we return to the plaques at the entrance of METESZ, where there is yet a fourth one entitled "Message of Geniuses to the Homeland and the Future." We present these messages in our own translation:

> With far-away future
> compare the present seriously.
> Have your impact; create; enrich;
> for our Homeland to shine.
>
> *Ferenc Kölcsey (1790–1838), author*
> *of the Hungarian National Anthem*

> The deepest source of
> all our progress, advancement,
> force, value, and even luck,
> is the educated mind.
>
> *István Széchenyi (1790–1860), reformer, founder of the*
> *Hungarian Academy of Sciences, "the Greatest Hungarian"*

> Not brute force;
> rather, it is culture that
> will make us stand out.
>
> *János Bolyai (1802–1860), mathematician,*
> *pioneer of non-Euclidean geometry*

Notice that the national feeling and dedicated patriotism here is not aimed against others. Kölcsey, Széchenyi, and Bolyai express what is as valid today as it was in their times, namely, that the best investment into the future of a nation is in education and science.

Notes

1 István Hargittai, "Albert Szent-Györgyi, Remembered." *Chemical Heritage* 2004, 22(2), 6–9.

2 Hargittai, "Albert Szent-Györgyi, Remembered," p. 7.

3 Hargittai, "Albert Szent-Györgyi, Remembered," p. 9.

4 Albert Szent-Györgyi, *Bioenergetics* (New York: Academic Press, 1957); Szent-Györgyi stated the same with slight variations on a number of occasions.

5 Paul Lendvai, *The Hungarians: A Thousand Years of Victory in Defeat* (translated from the German original by Ann Major; Princeton, NJ: Princeton University Press, 2003), p. 41.

6 Miklós Molnár, *A Concise History of Hungary* (translated from the French original by Anna Magyar; Cambridge, UK: Cambridge University Press, 2001), p. 30.

7 The "accents" as in Bárány do not represent accents, and there is no need to indicate where the stress is in the Hungarian language—the stress is always on the first syllable. Rather, what foreign authors often mistake for an accent is an indication of a different letter and a different sound. In Bárány, for example, both 'á's sound like the 'a' in "car."

8   From about the time of his Nobel Prize, Kertész then lived for many years in Berlin.

9   Arthur Koestler, *The Act of Creation* (New York: Macmillan, 1964).

10 Arthur Koestler, *The Sleepwalkers: A History of Man's Changing Vision of the Universe* (London: Hutchinson and Co., 1959).

11 Arthur Koestler, *The Watershed: A Biography of Johannes Kepler* (Garden City, NY: Doubleday & Co., 1960).

12 Arthur Koestler, *Darkness at Noon* (New York: Macmillan, 1941).

2

István Széchenyi

LEARNED SOCIETY

The statue of István Széchenyi on Széchenyi István
Square, District V, with four side figures; all by
József Engel (1880).

The teachings of István Széchenyi (1791–1860) sound timely even today. He had
innovative ideas and in a number of cases he turned his ideas into actual projects.
The expression "learned society" may refer to a science academy and it may also
refer to a well-informed society. Széchenyi recognized that Hungary needed a science academy, and with a circle of friends, he founded the Hungarian Academy
of Sciences (HAS).

The headquarters of the Hungarian Academy of Sciences, 9 Széchenyi István Square, District V.

The headquarters of the HAS is in a square in downtown Budapest, one of the most beautiful spots in the city. Since 2011, the square has carried Széchenyi's name. Previously, between 1947 and 2011, its name was Roosevelt Square, after Franklin D. Roosevelt; in honor of one of the leaders of the anti-Nazi coalition in World War II. Prior to that, it was Franz Joseph Square, named after the Emperor of Austria and King of Hungary who reigned between 1848 and 1916.

The building of the Hungarian Academy of Sciences is between Akadémia Street on the east and Széchenyi Quay on the west, on the side of the river. Behind the building, in the same block, on Arany János Street, is the Library of the HAS.

The plates (as of November 2012) inform about the name change from Roosevelt Square (after Franklin D. Roosevelt) to Széchenyi István Square.

Plaque dating from 1893 on the building of the Academy, commemorating the founding of the Hungarian Academy of Sciences in 1825.

A relief on the wall on the Akadémia Street side of the building commemorates the historic moment of the founding of the institution. This happened on November 3, 1825, at a session of the Hungarian Parliament in Pozsony (today, Bratislava, the capital of Slovakia). At the time, Pozsony was the seat of the Hungarian Parliament.

The relief depicts a scene in which Count István Széchenyi is offering one year's income of his estate to develop an institution for directing Hungarian spiritual life, especially for protecting and strengthening the Hungarian language. At the time, and through 1844, Latin was the official language of Hungary. There were three other principal founders of the Academy, Ábrahám Vay, Count György Andrássy, and Count György Károlyi, and they are seen along with other members of Parliament.

The relief on the wall of the Academy on the Akadémia Street side of the building (by Barnabás Holló, 1893). It commemorates the scene in which Count István Széchenyi is offering one-year's income of his estate to develop the Hungarian Academy of Sciences.

On August 18, 1827, King Francis I gave his royal assent to the law establishing the Hungarian Academy of Sciences. For forty years the Academy did not have its own headquarters, and the movement to build one started in the late 1850s. A broad-based collection brought together the necessary funds, and a spirited societal discussion ensued about the plans for building it.

Finally, the architect Friedrich August Stüler's design was accepted. Stüler was well known; by then he had designed the Nationalgalerie and the Neues Museum in Berlin and the Nationalmuseum in Stockholm. Miklós Ybl and Antal Szkalnitzky were in charge of construction.

Kornél Divald, an art historian member of HAS, compiled a book about the history of the building, which served as a basis for future studies.[1]

Above: Bust of Miklós Ybl on Margaret Island (by József Kampfl, 1975).

Left: Statue of architect Miklós Ybl (by Ede Mayer, 1896) at the foot of the Buda Castle, on Ybl Miklós Square, District I.

A plaque commemorates Kornél Divald, an art historian member of HAS, on the building where he lived and worked during the last two decades of his life, at 5 Budafoki Avenue, District XI.

The inauguration ceremony of the building took place on December 11, 1865. The building suffered heavy damage during World War II. Reconstruction was carried out in stages; the interior was completed in 1958.

The main facade, facing Széchenyi István Square, displays the coat of arms of Hungary and the inscription MAGYAR TUDOMÁNYOS AKADÉMIA (Hungarian Academy of Sciences). There are also the following wordings and years (in Roman numerals):

HAZAFIAK ALAPÍTOTTÁK MDCCCXXV (founded by patriots 1825)
MŰKÖDNI KEZDETT MDCCCXXXI (it began its operations 1831)
NEMZETI RÉSZVÉT EMELTE MDCCCLX (the nation erected it in 1860)
HÁZA FELÉPÜLT MDCCCLXIV (its home was completed in 1864).

The statues of six great scientists decorate the building: Isaac Newton and Mikhail Lomonosov on the western facade, facing the river; Galileo Galilei and Miklós Révai on the southern facade, facing the square; and René Descartes and Gottfried Leibniz on the eastern facade, facing Akadémia Street. Two of the statues, Newton and Leibniz, stand alone; Lomonosov and Galilei as well as Révai and Descartes stand in pairs. Four of the six statues are by Emil Wolff (1865), Révai is by Miklós Izsó (1865), and Lomonosov by Gyula Palotai Szkalos (1963).

Left: Isaac Newton, British scientist and mathematician. *Right, on the left*: The Russian scientist Mikhail V. Lomonosov (see text); *on the right*: Galileo Galilei, Italian astronomer, mathematician, and philosopher.

Left, on the left: Miklós Révai, founder of Hungarian linguistic historiography; *on the right*: René Descartes, French philosopher and mathematician. *Right*: Gottfried Leibniz, German philosopher and mathematician.

The original six were chosen in the middle of the nineteenth century, and four of them, Descartes, Galilei, Leibniz, and Newton, were giants of science. Originally, in Lomonosov's place, stood the statue of Rafaello Santi (better known as Raphael). The inclusion of an artist was justified because the Academy had a section for artists and writers (through the end of World War II). The inclusion of the Hungarian linguist, Miklós Révai, was in accordance with the aims of the Academy in protecting and strengthening the Hungarian language.

The Raphael statue was damaged during the war and those in charge of the Academy could have opted for its reconstruction in the early 1960s as part of the renovation of the building. However, they had other considerations. One was that the section for artists and writers had been removed from among the sections of the institution. The choice of Lomonosov was probably dictated by the fact that in 1961 there were celebrations for the 250th anniversary of his birth, and even more, that he was Russian and it was yet another gesture to strengthen Hungarian–Soviet ties. Lomonosov was a polymath, scientist, and writer, greatly revered in Russia as "our first university,"[2] but as a scientist, he was not of the same stature as Descartes, Galilei, Leibniz, or Newton.

The political implications of the choice emerge from the documents relating to it. As an obvious example of expediency, initially, the officers of the Academy proposed Vladimir I. Lenin, or Karl Marx for the replacement, and solicited instructions from the Hungarian Socialist Workers' Party (the communist party for all practical purposes in the one-party political system). This was too much even for the party and its Division of Culture and Science communicated its disapproval of the choice of Lenin or Marx for the replacement without giving any specific instructions in the matter. We know these details from an exchange between Ferenc Erdei, Secretary General of the Academy, and István Rusznyák, President of the Academy.[3] The scientist Mikhail Lomonosov thus turned out to be a reasonable solution.

In addition to the six scientists, ten statues decorate the facades of the building, four on its Danube side and six facing the Square. They symbolize arts and sciences which belonged to the original scope of the Academy.

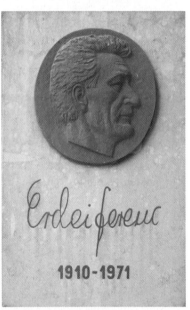

István Rusznyák's memorial plaque on the wall of the house where he used to have his apartment at 2 Füvészkert Street, District VIII. He was a professor of internal medicine and a long-time President of HAS (1949–1970).

Ferenc Erdei's relief by Árpád Somogyi under the arcades of the agricultural ministry at 11 Kossuth Square, District V (Chapter 5).

Four female statues on the facade of the Danube side symbolize Archeology, Poetry, Cosmography, and Politics.

Six female statues on the central section symbolize Law, History, Physical Geography, Mathematics, Philosophy, and Philology.

Statues of the linguist Gábor Szarvas and the historian Ferencz Salamon, both by Gyula Jankovits, in front of the Hungarian Academy of Sciences.

Statue of Ferenc Deák (by Adolf Huszár, 1887) on Széchenyi István Square.

Further to the south, but still on Széchenyi István Square, there is another big statue, commemorating the jurist statesman Ferenc Deák, who was also a member of HAS. He negotiated the Compromise of 1867 between the Habsburgs and the Hungarian nation. This deed opened the way for unprecedented prosperity in Hungary within the realm of the Habsburg Empire—the Austro-Hungarian Monarchy—up until the start of World War I in 1914. This period is called "the happy peace time," with considerable nostalgia. Deák earned the label "Sage of Fatherland."

Adjacent to Széchenyi Square on the south is the Eötvös József Square with a statue of József Eötvös, jurist, writer, and politician, a member of HAS, and father of the physicist Loránd Eötvös (more on them in Chapter 3).

Facing the Academy building from the Square, in front of its left edge, there is a slim black statue in a very different style from its environment.

Statue of historian Zoltán I. Tóth (by Tamás Vígh, 2006).

This commemorates the historian academician Zoltán I. Tóth (1911–1956). On October 25, 1956, he was a member of a delegation of instructors who wanted to inform the communist party leadership about the demands of the University. Near party headquarters, shooting broke out and he was killed by a stray bullet.

————— • ◆ • —————

The charter of the Hungarian Academy of Sciences gives it responsibility for overview of all scholarship in the country. The Academy comprises eleven sections: I, Language and Literature; II, Philosophy and History; III, Mathematics; IV, Agriculture; V, Medicine; VI, Technology; VII, Chemistry; VIII, Biology; IX, Economics and Law; X, Geosciences; and XI, Physics.[4]

The Academy consists of two principal components. One is the body of its members—no more than 365 academicians—among whom there cannot be more than 200 who are younger than 70 years of age. The membership is a two-tier system: there are corresponding members and ordinary members. One is first elected a corresponding member and later an ordinary member. All receive a life-time monthly stipend commensurate with a professorial salary. There are two additional membership categories, external members and honorary members. External members are Hungarians living abroad; either born there or expatriates. The honorary members are distinguished international scientists. In addition to the 365 academicians, the Academy comprises an extended body for all those who have a PhD degree or equivalent. The membership in this public body is voluntary and numbers well over ten thousand.

The PhD degree is earned from a university. There is also a higher doctorate, Doctor of the Hungarian Academy of Sciences, which can be earned from the Academy.[5] This broader membership elects its representatives according to their professional subdivisions and these representatives participate in much of the decision-making of the Academy except in the election of its new members. The elections constitute a triannual event and the number of new members is determined by the number of empty slots within the 365 limit.

In addition to the body of members, the other component of the Academy is the network of research institutes, which has its own governing body under the general direction of the Academy. The institute budgets have two principal sources: one is from the government through the Academy; the other is through various grants, mainly from the Hungarian Scientific Funds, the European Commission, and other institutions.

The Hungarian Academy of Sciences is supposed to be independent of politics; alas, in reality, complete political independence has seldom characterized its activities. Suffice it to mention two examples. Anti-Semitic considerations infected its election practices in the period between the two world wars, and for a quarter of a century, Jewish scientists and scientists with Jewish roots were seldom elected. After World War II, in 1949, with the start of communist dictatorship, the memberships of many bona fide scientists were downgraded to advisory roles—in practice, this meant exclusion from the Academy. In 1989, the Academy reinstated most of these memberships, but with rare exceptions this could only be done posthumously. The reinstatement was not automatic; thus, convicted war criminals were not included.

———— • ◆ • ————

Not far from the headquarters of the Hungarian Academy of Sciences, northbound along Széchenyi Quay, toward the Parliament building, there is a memorial plaque in Steindl Imre Street, honoring the architect Imre Steindl, who designed the Parliament building. Steindl was professor of architecture at the Budapest Technical University and a member of HAS.

Left: The plaque commemorates the architect Imre Steindl (by István János Nagy, 1967), at the corner of Széchenyi Quay and Steindl Imre Street, District V. *Right:* Steindl's bust in the assembly hall (Aula) of building "K" of Budapest University of Technology and Economics (see Chapter 6).

The Parliament building as viewed from Buda; Imre Steindl was its architect, Kossuth Lajos Square, District V.

One block further north, in the direction towards the Parliament, there is another Széchenyi statue, quite different from the one on Széchenyi Square, at the corner of Széchenyi Quay and Zoltán Street. This statue depicts a more urban and less aristocratic Széchenyi, holding a plan.

Statue of István Széchenyi (by József Somogyi, 1988), great reformer and the first Hungarian minister of transportation, at the corner of Széchenyi Quay and Zoltán Street.

István Széchenyi was an officer of the Imperial Army; in 1813, he participated in the war against Napoleon. From 1814, he traveled extensively, to Italy, France, and England, and also, to the Balkans and Asia Minor (Turkey). In 1825, he and his friend, Miklós Wesselényi, spent an extended period of time in England. Széchenyi recognized the backwardness of Hungary and attempted to convince his fellow landowners about the need for change. He did not speak perfect Hungarian, yet took the unusual step of giving a speech in Hungarian in the Hungarian Parliament, at a time when Latin was still the official language. In 1826, he resigned from the Army and from this point on he devoted his life to public service.

István Széchenyi helped to open the first casino and run the first horse races in Hungary. His first book in 1828 was about horse breeding. In 1830 he published his influential book about the lack of credit, which he found to be pivotal in hindering progress in Hungary. He argued for abolishing serfdom and for the introduction of wage-based employment; for equality for all before the law; and for terminating tax exemptions for the nobility. He stressed the importance of intellectuals in Hungarian society.

He did everything he could to facilitate modernization. He aimed at changing his homeland without disturbing its status as part of the Habsburg Empire. He involved himself in a plethora of activities, in founding new institutions and changing old ones. He encouraged building roads, making rivers navigable, and was instrumental in having the first permanent bridge built between Pest and Buda, the Chain Bridge (Chapter 6).

In a speech at the Hungarian Academy of Sciences, Széchenyi argued for the rights of nationalities and minorities and against mandatory spreading of the Hungarian language. He tried to create a centrist political entity. He was a great reformer and even his opponents respected him. At the time of their public debates, the politically more radical Lajos Kossuth, the popular leader and Governor of Hungary during the 1848–1849 War for Independence, called Széchenyi "the Greatest Hungarian." Széchenyi recognized that Kossuth's radical policies presented better opportunities for real reform.

Following the Revolution of March 15, 1848, and the declaration of Hungarian independence from the Habsburgs, Széchenyi accepted the appointment as minister of public works and transportation in the first independent Hungarian government headed by Lajos Batthyány. At the same time, he was appalled by the controversy between the Royal Court and the Hungarian reformers.

Széchenyi suffered a nervous breakdown on the eve of the armed conflict and on September 7, 1848, he was taken to a neurological nursing home in the Austrian Döbling, from which he never returned. In a few years' time, he recovered spiritually to the extent that he produced a few more books which made the Austrian government suspicious. They kept him under observation, harassed him, and the police confiscated some of his papers. Persecution and threats eventually drove him to suicide.

Ferenc Széchényi, István's father, founded the Hungarian National Museum. His statute (by János Istók, 1902) stands in the museum garden, 14–16 Múzeum Boulevard, District VIII.

Hungarian National Museum (Mihály Pollack's design) at 14–16 Múzeum Boulevard, District VIII.

János Arany's statue by Alajos Stróbl in front of the Museum is on the right.

István Széchenyi's father, Count Ferenc Széchényi (1754–1820), founded the Hungarian National Museum.[6]

János Arany (1817–1882) was a poet and translator. He started his career as a teacher, but continued his education on his own, which included learning the English language. He became a national poet and published studies on esthetics and literary history. Among his translations, we would mention translation of the complete works of Shakespeare into Hungarian. He was a member of HAS and was for a long time (1865–1879) its secretary general.

Grave of János Arany
at the Fiume Avenue
National Necropolis.

Bust of János Arany at the Hungarian
Academy of Sciences.

There are other memorabilia in the garden of the Hungarian National Museum, relevant to our discussion. Flóris Rómer (1815–1889) was a Benedictine monk, art historian, and archeologist. He spent years in incarceration following the supression of the War for Independence of 1848–1849. Later, he worked at the Hungarian National Museum and was active in domestic and international interactions among archeologists, art historians, and related professionals. Ferenc Pulszky (1814–1897) was an art historian, archeologist, and politician, with a law degree. Following the supression of the War for Independence of 1848–1849, for almost two decades he lived in exile. After the Compromise, he was active in museum administration. József Hampel (1849–1913) was an archeologist with a law degree. He worked at the Hungarian National Museum and taught archeology at the University of Budapest. All three were members of HAS.

Memorial wall honoring three archeologists in the garden of the Hungarian National Museum (by József Damkó, 1916). *From left to right*: Flóris Rómer, Ferenc Pulszky, and József Hampel.

Ferenc Kazinczy (1759–1831) was a writer, poet, and translator. His fame is primarily due to his activities in the reform of the Hungarian language. Kazinczy and the other participants in language reform coined or renewed hundreds of words. These reforms modernized the language and made it possible for it to become the official language of the land in 1844. Kazinczy engaged in politics when he joined the Hungarian Jacobian conspiracy.[7] In 1795, he was arrested, tried, and sentenced to death, but his death sentence was commuted, and he was freed after years of incarceration. He was a member of HAS and had participated in its founding.

Ferenc Kazinczy's bust (by Miklós Vay, 1861) stands in the garden of the Hungarian National Museum. The memorial plaque at 1 Kazinczy Ferenc Street, District VII, says that Kazinczy was an apostle of the reformation of the Hungarian language; he participated in the Hungarian Jacobian movement, and he spent 2387 days in prison.

Marble statue honoring Mihály Vörösmarty and his Szózat (a call to the nation) on Vörösmarty Square, District V. It is the work of Ede Kallós and Ede Telcs, unveiled in 1908. The stand carries an inscription of the first words of Szózat, "Hazádnak rendületlenül/Légy híve, oh magyar" ("Be forever faithful to your Homeland, Oh Magyar!" Translation from Bob Dent, *Every Statue Tells a Story* (Budapest: Európa Könyvkiadó, 2009), p. 138).

Walking from Széchenyi István Square southward through Dorottya Street we reach Vörösmarty Mihály Square named after the great poet and a member of HAS. In the center of the square there is his magnificent marble statue.

Mihály Vörösmarty (1800–1855) studied law and became a lawyer. He participated in the reform movement and, as Lajos Kossuth's follower, in the struggles in 1848–1849. His *Szózat* is considered to be the second national anthem of Hungary. He wrote it in 1836 and the music was composed by Béni Egressy in 1840. Vörösmarty was the favorite poet of the Hungarian–American Nobel laureate physicist Eugene P. Wigner. Ever since Wigner had learnt about Vörösmarty in high school, he had followed the poet's teaching about the need to leave behind a legacy. Wigner often quoted the last three lines of one of Vörösmarty's poems; in Wigner's own translation: "The world won't last forever/But while it lives and while it lasts/it builds or rents but never rests."[8]

———— • ❖ • ————

The predecessors of the Hungarian Natural History Museum developed within the framework of the Hungarian National Museum. It became independent in 1963 and acquired its present name in 1991. From the mid-1990s, its principal collection has been at 2–6 Ludovika Square, District VIII, but some of its collections are still scattered among other locations in the city.[9]

Entrance to the central building of the Hungarian Natural History Museum and its lizard decoration at 2-6 Ludovika Square, District VIII.

The Museum consists of the Departments of Anthropology, Botany, Geology and Paleontology, Mineralogy and Petrology, Zoology, and the Molecular Taxonomic Laboratory.

Great travelers have contributed much to linguistics and oriental historiography, and some have memorials in Budapest, ranging from a simple memorial plaque to sometimes several statues. Sándor Kőrösi Csoma (1784–1842) was from Transsylvania. Early on in his youth he acquired the romantic notion that he should dedicate his life to finding the origin of the Hungarians. He became a polyglot both in western and eastern languages, studied in Hungary and Germany, and prepared himself in all possible ways for several trips into the unknown East.

In 1819, he embarked on his first trip through the Balkans, Egypt, the Near East, and Persia, to Central Asia. He walked most of the 12,000-kilometer (7500-mile) distance. He lived a spartan existence and often declined financial assistance. Under British influence, he became interested in Tibetan language and culture and made Tibetology his side project—in hindsight, it became his principal contribution to world culture. As fruits of his scholarship, he published the first Tibetan–English dictionary and a book about the Tibetan language. In India, he worked as a museologist.

In 1842, he began what turned out to be his last trip. His goal was to reach Lhassa, the Tibetan capital, but somewhere in India he contracted malaria and succumbed to the illness. He left behind a treasure trove of books and documents. He is buried in Darjeeling and his grave has become a place of pilgrimage. He was a member of HAS and other learned societies.

Ármin Vámbéry (1832–1913) was another polyglot and great traveler, a scholar of the East, especially the Turkish language and the Ottoman Empire. He suffered from several handicaps; he was Jewish, poor, lost his father when he was one year old, and walked on crutches due to a congenital disorder. Yet he was gifted, adventurous, and capable of gaining the trust and support of influential people. He spent years in Turkey learning the language and culture, published

Sándor Kőrösi Csoma: The statue (by Géza Csorba, 1968) stands in the garden of the Ráth György Museum (of oriental and Indian art), at 12 Városligeti Alley, District VI.

The bust (by Barnabás Holló, 1909) at the Hungarian Academy of Sciences.

Statue of Sándor Kőrösi Csoma (by Béla Tóth, 1984) on Kőrösi Csoma Walkway, District X.

a German–Turkish dictionary, and eventually became professor of eastern languages at Budapest University.

In 1863–1864, he was on an especially daring journey. Disguised as a Sunni dervish, he joined a caravan headed for Central Asia where Vámbéry collected unique information from a Western point of view, but could not keep proper records without inviting suspicion. Later he reconstructed and published his experiences. He augmented his Central Asia tour with a side trip to Persia.

The literature about him projects him as a double agent and a double dealer—not very attractive traits—but nobody has questioned his intelligence and resourcefulness. A picture emerges of him as a dedicated scholar and a useful and appreciated agent of the British Empire. His theories about the close relationship between the Hungarian and Turkish languages have been discredited in Hungary, but they have gained respect in Turkey. Vámbéry was a member of HAS.

Manó Kogutowicz (1851–1908) was a geographer and cartographer. He came from southern Poland and received training at the Military Technical Academy in Vienna. He excelled in the production of maps. He was close to thirty years old when he and his future wife moved to Hungary. He taught in Sopron and, as a side project, he received commissions to draw maps. In 1883, the Ministry of Religion and Public Education invited Kogutowicz to Budapest to head an institute of cartography. From then onwards he was engaged full time in his favorite occupation. Eventually, he founded a company that later became the Hungarian Institute of Geography, and for a long time the Hungarian school system used its maps almost exclusively.

Plaque commemorating Ármin Vámbéry on the house where he lived at 24 Belgrád Quay, District V.

Memorial plaque for Manó Kogutowicz at 8 Széchenyi Quay, District V (by Béla Domonkos), where between 1892 and 1913 the Hungarian Institute of Geography operated.

Ignác Kúnos (1860–1945) was a linguist and Turkologist. He was Ármin Vámbéry's pupil, taught at Budapest University as Professor of Turkish Philology, and was a member of HAS. For years, he worked as a guest professor in Turkey.

Jenő Cholnoky (1870–1950) was a geographer and engineer. His works extended to geography, hydrology, and climate studies. He was a prolific author and he popularized science through his books and lectures. He was a member of HAS.

Gyula Prinz (1882–1973) was a geographer, geologist, and ethnographer; he taught first at the University of Pécs, later at the University of Szeged. He

Tombstone of Ignác Kúnos in the Kozma Street Jewish Cemetery (plot 5B-10-22), District X.

participated in expeditions to Central Asia. His activities embraced the totality of geography. He preserved his liberal demeanor even in periods of political extremism. He was a member of HAS, but in 1949, his membership was downgraded and it was only reinstated, posthumously, in 1989.

Frigyes Pesty (1823–1889) was a historian, a member of HAS, and his principal interest was in historical geography. He collected the names of towns and villages from all over Hungary.

Left: Jenő Cholnoky's memorial plaque (bronze relief by Béla Domonkos) at 1 Gyulai Pál Street, District VIII. *Right*: Gyula Prinz's memorial plaque at 10 Pauler Street, District I.

Left: Memorial plaque of Frigyes Pesty (by Antal Czinder, 1997), 20 Váci Street, District V. *Right*: Memorial plaque of István Györffy (by Árpád Somogyi) on the wall of the Museum of Ethnography, Kossuth Lajos Square, District V.

István Györffy (1884–1939) was an ethnographer and a member of HAS. He was the first professor of ethnography at the University of Budapest. His principal interest was in the origins and development of various Hungarian ethnographic groups.

Gyula Germanus (1884–1979) was a resourceful youth. When his parents could not afford to buy a piano, he *drew* a keyboard and practised on it. He was deeply interested in Eastern history, philosophy, and religions. He learned Turkish and other Eastern languages. Ármin Vámbéry tutored him and encouraged his interests. Germanus traveled to Turkey and participated in revolutionary political movements. He was almost executed for his participation, but later, when his political comrades took over Turkey, he became a much respected friend of the new leaders of the country.

Gyula Germanus's bust (by György Szabó, 2001) in the Germanus Gyula Park at the Buda bridgehead of Margaret Bridge, District II.

Germanus taught in Budapest, served as an official envoy for Hungary in international interactions, and worked as a war correspondent during World War I. At one point, he formed and helped to spread unscientific, nationalistic, and racist views advanced by some politicians, including the anti-Semitic Pál Teleki. Later, Germanus's own scholarship disproved these views. A park is named after him in District II, and there is a memorial plaque on the wall at 3–5 Petőfi Square, District V, where Germanus spent the last two decades of his life.

After Turkey, Germanus's interest turned to India, and he spent years in Bengal, a region today divided between India and Bangladesh. His interest in and deep knowledge of Islam led to his conversion to the Muslim religion. He made extensive trips and pilgrimages to Arabic countries. He was a prolific author and between the two world wars his successful books shared his experiences with a growing audience. During the anti-Semitic frenzy in Hungary, Germanus was also obliged to wear the yellow star because of his Jewish origin. He remained active in teaching and writing into his eighties.

Ervin Baktay (1890–1963) was a painter, art historian, orientalist, writer, and translator. On one of his visits to India, he spent years visiting the places where Sándor Kőrösi Csoma had been before him. He wrote books, translated others, taught, ran exhibitions, and gave lectures, especially to spread information about

Pál Péter Domokos's plaque by Gergely Csoma at 10/c Bartók Béla Avenue, District XI.

the culture of India. There is a memorial plaque for Ervin Baktay on the house where he lived, at 16/a Eszék Street, District XI.

Pál Péter Domokos (1901–1992) was a historian and ethnographer, especially interested in the history of the Csángó: Hungarians living in Moldova. He was from Transylvania, but moved to Budapest. In the 1950s, the secret police prevented him from doing research and teaching, but eventually he received

Memorial plaque of Lajos Ligeti, orientalist, at 26 Belgrád Quay, District V.

Plaque of László Kardos (by Iván Szabó), ethnographer, at 2 Semmelweis Street, District V.

Memorial plaques for (*left*) István Mándoky Kongur (by Sándor Győrfi), at 52 Bartók Béla Avenue, District XI, and (*right*) for László Országh (by Antal Czinder), at 12 Balaton Street, District V.

assignments in gimnáziums. He amassed a huge ethnographic collection and founded a fellowship for Csángó youth who would be willing to return to their homeland after graduation.

Lajos Ligeti (1902–1987) was an orientalist, who made extensive trips to the East, including Mongolia, Afghanistan, and Japan. As with several of his predecessors, he was deeply interested in the origins of the Hungarians. He was vice president of HAS for many years and held positions in international science organizations.

László Kardos (1918–1980) was an ethnographer, who was sentenced to heavy inprisonment for his participation in the 1956 Revolution.

István Mándoky Kongur (1944–1992) was an orientalist with a special flair for languages, and he benefited from his interactions with a Soviet garrison where he found soldiers who spoke a plethora of minority languages. He made extensive trips to Soviet Central Asia, Mongolia, and elsewhere. He was of Cumanian origin and researched the Cumanian language.

László Országh (1907–1984) was a professor of English language and literature at Debrecen University. He is most famous for his English–Hungarian and Hungarian–English dictionaries. He founded American studies in Hungary.

———— ◆ ————

The reform movement in Hungary during the first half of the nineteenth century suffered from many handicaps. One of these was insufficient information about the country. Elek Fényes (1807–1876), an economist, statistician, and writer, realized this when in the 1830s he participated in the activities of the Hungarian Parliament. Fényes embarked on writing a monumental work to describe the country. Once in possession of information about Hungary, he became a yet stronger advocate of reform, including the emancipation of serfs. Following the Revolution of March 15, 1848, the first independent Hungarian government appointed Fényes to be in charge of the newly organized national office of statistics. The suppression of the War for Independence brought trying times for Fényes. He was imprisoned for some time and his career fell apart, but he continued his studies in statistics. He was a member of HAS.

The Central Office of Statistics (Központi Statisztikai Hivatal, KSH) and its predecessors have existed since 1867. Its headquarters are at 5–7 Keleti Károly Street, District II. In 1896, architect Győző Czigler designed the building and it opened for business in 1897. Károly Keleti (1833–1892) was the organizer and first director of the Central Office of Statistics. He was a member of HAS.

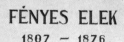

FÉNYES ELEK
1807 — 1876

STATISZTIKUS ÉS FÖLDRAJZ TUDÓS
KOSSUTH HIVE ÉS MUNKATÁRSA,
AZ 1848-AS STATISZTIKAI HIVATAL
VEZETŐJE, A MAGYAR TUDOMÁNYOS
AKADÉMIA TAGJA.

Elek Fényes's memorial plaque at the corner of Fényes Elek and Keleti Károly Streets, District II.

Central Office of Statistics at 5–7 Keleti Károly Street, District II. Győző Czigler was its architect.

Plaque commemorating Károly Keleti at 18/a Keleti Károly Street where the Central Office of Statistics stands.

József Kőrösy's memorial plaque at 30 Délibáb Street, District VI (the corner of Délibáb Street and Dózsa György Avenue).

József Kőrösy's (1844–1906) immediate interest was public health and demography. He was an internationally renowned statistician and the founder and first director of the Budapest office of statistics. He was a member of HAS. His grandson, Ferenc Kőrösy, was a chemical engineer and authority on applied physical chemistry, an external member of HAS, and a leading Israeli scientist (Chapter 8).

———— ◆ ————

The Hungarian Academy of Sciences is an all-inclusive learned society, including musicology. Hence we mention here some of the statues of Béla Bartók, Zoltán Kodály, and Ferenc Liszt.

Béla Bartók (1881–1945) was a composer, pianist, and musicologist. His interest turned toward folk music after 1905 and he and Zoltán Kodály (1882–1967) embarked on an ambitious project. They toured the countryside studying indigenous music. Kodály limited his collecting activities to the Hungarian-speaking territories. Bartók had early on crossed not only national but also language frontiers, and extended his activities as far as North Africa and the Arab lands.

Bartók and Kodály collected folk music, probably at the last moment that genuine music could still be gathered from authentic practitioners. The Hungarian

Left: Statue of Béla Bartók on Kosztolányi Dezső Square, District XI; the art of József Somogyi from 1981. Bartók stands at a gate displaying eleven bells of different sizes and shapes. *Right*: Béla Bartók's bust by András Beck on the Művész Walkway, Margaret Island.

Left: Bartók's statue in the garden of the Bartók memorial house, 29 Csalán Avenue, District II. Imre Varga's creation was unveiled in 1981, during the Bartók Centennial. *Right*: Béla Bartók's statue by András Beck (from 1954, unveiled on this location in 2008) in the garden of the Institute for Musicology of HAS, at 7 Táncsics Mihály Street, District I.

Academy of Sciences recognized the value of their studies and discoveries by electing them to its membership and providing support for their research.

Both Bartók and Kodály were anti-fascist and resented Hungary's strengthening ties with Nazi Germany. They were among a small group of intellectuals who protested the harsh anti-Jewish laws of Hungary. Bartók and his wife left Hungary for New York in 1940. He died in New York shortly after the conclusion of World War II. In 1988, his ashes were brought back to Budapest and interred in Farkasréti Cemetery.

Kodály was a composer, musicologist, and a great pedagogue. His pedagogical methods have spread all over the world and are better known than his music. He

Two images of Kodály's statue in Püspök Garden in the Castle District, District I. The statue by Imre Varga was unveiled in 1982 elsewhere in the Castle District, but it has been in its present location since 2003.

Zoltán Kodály's memorial plaque on the wall of the house where he used to live (see next figure) and where there is a Kodály Memorial Museum and Archives today. Kodály is depicted with his gramophone on this plaque (by Tamás Vígh, 1972).

1 Kodály Circle, District VI, where Zoltán Kodály lived.

Béla Bartók's tombstone (by Miklós Borsos, plot 60/1-főút-9/12) is on the left and Zoltán Kodály's (by Pál Pátzay, plot 20, körönd-1-17/18) is on the right, both in the Farkasréti Cemetery.

wielded great respect and he was President of the Hungarian Academy of Sciences between 1946 and 1949. Kodály died in Budapest and his remains rest in the Farkasréti Cemetery.

The Liszt Ferenc Academy of Music is a prestigious institution of musicology. Its headquarters are at 8 Liszt Ferenc Square, District VI; its renovation was completed in 2013.

Central part of the main building of the Liszt Ferenc Academy of Music, at 8 Liszt Ferenc Square, District VI. The statue of Ferenc Liszt (1811–1886) is by Alajos Stróbl (1907).

Statue of Ferenc Liszt at the Hungarian State Opera House by Alajos Stróbl (1885), at 22 Andrássy Avenue, District VI.

Bust of Ferenc Liszt on Margaret Island, close to the water tower, by Dezső Erdey (1960).

Statue of Ferenc Liszt playing an invisible piano by László Marton (1986), at Liszt Ferenc Square, District VI.

We conclude this section by presenting the memorial plaques honoring two world-renowned musicologists. Bence Szabolcsi (1899–1973) was a scholar of music and music history and professor of the Liszt Ferenc Academy of Music. His publications cover all aspects of Hungarian music. He founded the Bartók Archives from which the Institute of Musicology of the Hungarian Academy of Sciences has developed. Dénes Bartha (1908–1993) specialized

Left: Plaque of Bence Szabolcsi at 40 Pozsonyi Avenue, District XIII. *Right*: Plaque of Dénes Bartha (by László Dinyés, 2008) at 87 Attila Avenue, District I.

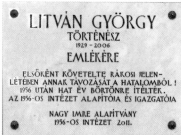

Memorial plaques at 40 Pozsonyi Avenue, District XIII, for Lajos Hatvany (*left*), author and historian of literature, and György Litván (*right*), historian of 1956 and twentieth-century Hungary.

in researching Haydn's oeuvre. He joined Szabolcsi in some of his publishing activities. Bartha taught in Budapest and, for about a decade and a half, at Harvard and other US universities. Both Szabolcsi and Bartha were members of HAS.

———•◆•———

Szabolcsi's memorial plaque is one among several on the same residential building. We mention here two more.

Lajos Hatvany (1880–1961) was a representative of progressive spirit and played an active role in the cultural life of Budapest. He had to spend much of his career abroad between the two world wars and during World War II. György Litván (1929–2006) stood up to the communist dictator Mátyás Rákosi prior to the 1956 revolution and continued as an outspoken critic during the Kádár regime that followed. After the political changes he became the first director of the 1956 Institute, founded to research the history of the 1956 revolution and twentieth-century Hungary.

Notes

1 Kornél Divald, *A Magyar Tudományos Akadémia palotája és gyüjteményei: Kalauz az Akadémia megbízásából.* (The Palace and Collections of the Hungarian Academy of Sciences: Compendium Commissioned by the Academy; Budapest: Magyar Tudományos Akadémia, 1917); György Rózsa, *A Magyar Tudományos Akadémia palotája* (The Palace of the Hungarian Academy of Sciences; Budapest: A Magyar Tudományos Akadémia Könyvtára, 1982).

2 See, e.g. P. N. Nikolaev, "Mikhail Vasil'evich Lomonosov, Our First University." *Physics-Uspekhi* 2011, 54, 1155–1160.

3 We are grateful to Diana Hay, director of the Archives of the Hungarian Academy of Sciences, for the relevant documentation.

4 It appears odd that Physics is Section XI, but for a long time Section III comprised of mathematics and physics together. They split when the system of sections had already been established a long time before, and it was impractical to renumber them.

5 In Anglo-Saxon countries, this higher doctorate does not exist or if it does, it has little significance in recognition and promotion. This higher doctorate is only similar but not the same as habilitation in the German system. Habilitation also exists in Hungary. Whereas in acquiring the degree of Doctor of the Hungarian Academy of Sciences, the emphasis is on research achievements; habilitation is granted by universities and, in earning it, the emphasis is on teaching.

6 There is a slight difference in the way father and son spelled their last name.

7 So named after the Jacobian Club, the most famous group of French revolutionaries.

8 See, e.g. at least in three places in *The Collected Works of Eugene Paul Wigner*, Volume VII, *Historical and Biographical Reflections and Syntheses*, Annotated and edited by Jagdish Mehra (Berlin and Heidelberg: Springer, 2001), p. 319; p. 392; p. 463. The Hungarian original, Mihály Vörösmarty (1843), the last three lines from "Keserű pohár, [A bitter cup]. "Örökké a világ sem áll;/De amig áll, és amig él,/Ront vagy javít, de nem henyél."

9 The website of the museum: http://www.nhmus.hu/hu.

3

Loránd Eötvös

UNIVERSITAS

The bust of Loránd Eötvös (by László Csejdy, 1991) in front of building "A" of Eötvös University's Trefort Garden (4 Múzeum Boulevard, District VIII).

Loránd Eötvös (1848–1919) was a world-renowned physicist who was instrumental in raising the level of Hungarian science and education. The "science" university today carries his name, but it is only part of what "universitas" means, because some of the areas of higher learning belong to other schools.

King Sigismund (Zsigmond, 1368–1437) founded a university in 1395 in Óbuda, but it was short-lived. Everything that became permanent began in 1635, when Péter Pázmány (1570–1637), the Archbishop of Esztergom, decided to found a university.[1]

It was under the Jesuit Order, and Pázmány secured the approval of the Habsburg Emperor, Ferdinand II, though failed to secure recognition from Pope Urban VIII. At the time, much of Hungary was still under Turkish occupation, and instructions started in Nagyszombat (now, Trnava, Slovakia) in the Faculty of Humanities, followed by the Faculty of Theology. The Faculty of Law became the

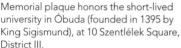

Memorial plaque honors the short-lived university in Óbuda (founded in 1395 by King Sigismund), at 10 Szentlélek Square, District III.

Statue of Péter Pázmány (by Béla Radnai, 1914) on Horváth Mihály Square, District VIII.

third, and the Faculty of Medicine, the fourth. For a long time, these four faculties comprised the University.

In 1777, Maria Theresa (1717–1780), Queen of Hungary, Empress of Austria, and sovereign of many other lands, decided to move the university from Nagyszombat to Buda, which was the intellectual center of Hungary. The country was still in the process of recovering following the long Turkish occupation and the university was an issue of concern for the whole country rather than just for the city.

In 1780, instructions started in Buda. Maria Theresa's son and successor, the enlightened King Joseph II (1741–1790), intended to modernize higher education. Jozsef II was the Emperor of the Habsburg Empire and the Hungarian King between 1780 and 1790. He was an unusual sovereign who did not allow himself to be crowned—hence his nickname: the king with a hat. He was an enlightened absolutist and introduced numerous reforms. He improved education, and established the engineering school, later known as *Institutum Geometricum*, as part of the university. In 1783, Joseph II decided to move the University from Buda to Pest to give new vitality to its development. There is a memorial plaque on the wall of the Royal Palace, which serves currently as the central building of the Hungarian National Gallery, at 2 Szent György Square, District I. The plaque commemorates the operations of the University of Buda between 1777 and 1784. In 1786, another new entity came to life for the training of veterinarians, following the example of Vienna University, the *Institutum Veterinarium*.

Statues of Queen Maria Theresa and King Joseph II on the southern side of the western facade of the Parliament building.

The Hungarian language was gaining ground in university instruction during the era of reforms, between 1825 and 1848. The protection and strengthening of the Hungarian language was a central aim of the newly established Hungarian Academy of Sciences. The aspirations of the University and the Academy reinforced each other. The introduction of the Hungarian language into the sciences though was far from trivial.

The Benedictan monk Ányos Jedlik (1800–1895) was a physicist, inventor, and a member of HAS. He studied in Hungary and taught several science subjects in a variety of schools. He played a role in the reform movement, but he also carried out menial work when the need arose to dig trenches during the War for Independence in 1848–1849. He was professor of physics between 1840 and 1878 at the University of Budapest. He invented a machine for the production of soda-water, did research on electricity, and invented the principle of the dynamo for generating electricity. He contributed to the creation of the Hungarian scientific language and was the first to lecture on physics in Hungarian. He served at one time as Rector of the University.

Busts of Ányos Jedlik, *from left to right*: at the spherical aula (see below) of Eötvös University (by István Szabó, Jr., 1999); at the Museum of Transportation (Chapter 6); and in the Museum of Electro-techniques (Chapter 6).

The Revolution and War for Independence of 1848–1849 brought about reforms in university life, but the changes were aborted after the defeat of Hungary. Loránd Eötvös's father, József Eötvös (1813–1871), returned to these reforms many years later. He was one of the most influential politicians of culture and education in the nineteenth century. In 1848, he served as minister of religion and public education in the first independent Hungarian government. He aimed at making the university into a secular institution. Already at this time, he raised the need to separate the faculties of humanities and the sciences, something that would be accomplished only a century later. He was minister of religion and public education again, after the Compromise between Austria and Hungary, from 1867 until his death. He was a member of HAS and, toward the end of his life, its president. He was also a significant writer.

The statue of József Eötvös by the foundryman turned sculptor, Adolf Huszár. It stands on Eötvös József Square, District V.

The University Library building, opened in 1876, at 6 Ferenciek Square, District V.

In the 1850s, the Faculty of Philosophy, which comprised philosophy and various humanities, expanded. Several areas of science, such as mineralogy, botany, zoology, and chemistry left the medical faculty and joined the Faculty of Philosophy. On the other hand, engineering left the Faculty of Philosophy to form an independent entity (Chapter 6). The Faculty of Medicine made great strides to become a modern institution with superb professors, including Ignác Semmelweis in obstetrics (Chapter 4).

The 1867 Compromise between the Habsburgs and Hungary opened a new era not only in the life of the University but also in the development of the whole

Left: Statue of Ágoston Trefort (by József Kiss, 1904) in Trefort Garden, between Múzeum Boulevard and Puskin Street, District VIII. *Right*: Ágoston Trefort's bust in the entrance hall of the Library of the Budapest University of Technology and Economics.

country. In 1873, Buda, Pest, and Óbuda united to form Budapest. In the period between the Compromise and the outbreak of World War I, 1867–1914, there was unprecedented progress in virtually all aspects of the life of the nation. The University continued to consist of four faculties, theology, philosophy, law, and medicine. Strong high schools joined the University to improve the training of teachers by providing them with the opportunity to do practice teaching (Chapter 7).

Ágoston Trefort (1817–1888) was a politician and political writer. In his youth, he traveled extensively in Western Europe and wrote studies on historiography and economics. In 1841, the Hungarian Academy of Sciences elected him to its membership. In 1848, he participated in the first independent Hungarian government. After the suppression of the War for Independence, he withdrew from public life. In the 1860s he became active again and assumed important positions following the Compromise. From 1872, he served as minister of religion and public education and from 1885, also as President of the Hungarian Academy of Sciences.

The Nobel laureate James D. Watson and Magdolna Hargittai in front of a memorial plaque for Ágoston Trefort in the former Mintagimnázium (see Chapter 7), today, Trefort Gimnázium, 8 Trefort Street, District VIII.

The Rector's Office of the University and the Faculty of Law, at 1-3 Egyetem Square, District V.

There was unprecedented development in Hungarian higher education during Trefort's years in office. He had an array of new buildings constructed for the University, including the rector's office, the University Library, departments, clinics, and laboratories. He helped the faculties, especially medicine and the natural sciences to acquire new institutions and employ professors who were dedicated to research in addition to teaching.

He was instrumental in the secularization of university education and promoting equality among the various religions, including the recently emancipated Judaism. There was a noticeable increase in the number of Lutheran students between 1867 and 1914, to one and a half times their number only a few years before, thus indicating the interest of the German minority in the intellectual professions. The number of Jewish students doubled, reaching one third of the total student population from one fifth a few years before.

In contrast, women were still barred from university education at Trefort's time. He opposed the emancipation of women and declined their admittance. He showed rigidity in a compelling case when he barred a foreign-educated female physician from practicing her profession in Hungary. Vilma Hugonnai (1847–1922) was born into an aristocratic family (the original spelling of her name was Hugonnay—the "y" ending reflected nobility—but she preferred using it as Hugonnai). She wanted to obtain higher education, but that was impossible for women in Hungary when she was young. In 1869, she left for the medical school of Zurich University where they trained women physicians. In 1879, the University of Zurich conferred the medical doctor degree on her, and after one additional year gaining practice in surgery, she returned to Hungary.

At home, she could not get her diploma registered. Trefort suggested to her that she gain the necessary qualifications to be a midwife, and for years she worked in this capacity. Eventually, the Hungarian regulations changed, and she finally received permission to study for her Hungarian physician's degree. In 1897, her dream came true and she worked as physician for the rest of her life. A street is named after her in Nagytétény, District XXII, where she was born. The district erected a bust in her memory at the meeting of Hugonnay Vilma Street and Kastélypark Street, but it was stolen. There is a memorial plaque on

Above: Grave of Vilma Hugonnai and her husband, Vince Wartha, and inscription at the Fiume Avenue National Necropolis. Wartha was a chemist member of HAS.

Left: The first female graduate from a Hungarian medical school Sarolta Steinberger's tombstone (plot 5B-7-10) in the Kozma Street Jewish Cemetery, District X.

the house where she lived, at 41 Bíró Lajos Street, District VIII. Sarolta Steinberger (1875–1966) was the first female physician to graduate from a Hungarian medical school.

Mrs. Pál Veres, née Hermina Beniczky (1815–1895), founded a women's school of higher education. According to the inscription on her statue, she fought to enable women to become contributors to the prosperity of the nation by her education and by her heart.

The area between Múzeum Boulevard and today's Puskin Street (formerly, Eszterházy Street) has been called Trefort Garden since 1904 when the Trefort statue was erected there. The building at 14–16 Puskin Street used to belong to the Hungarian Science Association. Today, it is rather neglected, but there is still a Charles Darwin stone relief on its wall.

The statute of Mrs. Pál Veres (by György Kiss, 1906) has been in its present location since 2007, at the beginning of Veres Pálné Street, District V.

A relief of Charles Darwin (by Szilárd Sződy, 1933) on the wall of the former offices of the Hungarian Science Association at 14–16 Puskin Street, District VIII.

Trefort Garden is a small university campus, which a variety of schools have shared at one time or another. The development of the physical plant started under József Eötvös. The first structure was erected for chemistry in 1871 with sophisticated laboratories that cost much more than just an ordinary building with offices and lecture rooms. Today it is Building "B."

Reliefs of portraits of great scientists decorate the interior of Building "B" although the building now serves departments in the humanities. At the time

Building "B," at 4/b Múzeum Boulevard, District VIII; originally it housed chemistry; today it accommodates departments in the humanities.

when this chemistry building started its operation, some of the reliefs depicted quite recent scientists. Now they are all historical.

The main building of Trefort Garden, at 6–8 Múzeum Boulevard, was designed by Imre Steindl. Originally it was for the Technical University, but in 1909 the University of Budapest took it over when the Technical University acquired its magnificent central building in Buda (Chapter 6).

Henry Cavendish (1731–1810), English natural philosopher and chemist; John Dalton (1766–1844), English natural philosopher and chemist; Joseph Priestley (1733–1804), English clergyman and chemist; and Carl Wilhelm Scheele (1742–1786), Swedish chemist.

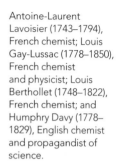

Antoine-Laurent Lavoisier (1743–1794), French chemist; Louis Gay-Lussac (1778–1850), French chemist and physicist; Louis Berthollet (1748–1822), French chemist; and Humphry Davy (1778–1829), English chemist and propagandist of science.

Jöns Jakob Berzelius (1779–1848), Swedish chemist; Leopold Gmelin (1788–1853), German chemist and physiologist; Eilhard Mitscherlich (1794–1863), German chemist; and Heinrich Rose (1795–1854), German mineralogist and chemist. The double portrait displays the French chemists Auguste Laurent (1807–1853) and Charles Gerhardt (1816–1856).

Upstairs in Building "B," there is a bust of Sándor Kőrösi Csoma (Chapter 2) symbolizing the returning departments of the humanities.

In 1886, two important additions enriched the campus. One of these was Building "A" at 4 Múzeum Boulevard which housed the zoological, botanical, mineralogical, and other collections. The enormous hall built especially for the spectacular mineralogical collection was breathtaking even when we visited it a century later. The other addition was Building "D" for physics which became famous for Loránd Eötvös's seminal experiments.

Loránd Eötvös graduated from the Gimnázium of the Piarist Order. He studied law at the University of Pest, then physics in Heidelberg, Germany, where he acquired his doctorate. He had great science teachers, such as G. Kirchhoff, H. Helmholtz, and R. Bunsen. Loránd Eötvös was elected at the age of twenty-five to membership of the Hungarian Academy of Sciences, which he later served

Left: The main building of Trefort Garden, at 6–8 Múzeum Boulevard, District VIII. In 1950, the humanities and the natural sciences split into two separate faculties. For half a century, Trefort Garden housed the natural sciences. By now, the natural sciences had moved to their new campus, and the humanities returned to Trefort Garden. *Right*: Building "A" at Trefort Garden, now part of the Faculty of Humanities. It is at 4 Múzeum Boulevard, District VIII.

Building "D" used to
house Loránd Eötvös's
physics laboratory.

as president for many years (1889–1905). He was professor of physics, and his
investigation of gravitation meant an experimental contribution to Albert Einstein's general theory of relativity.

The headquarters of the Hungarian Geological Institute is in a beautiful secessionist (Art Nouveau) building designed by Ödön Lechner (Chapter 4).

Left: The plaque (by István Kákonyi, 1959) on the wall of the Puskin Street side of
Building "D", which states that Loránd Eötvös lived and conducted his seminal
research in this building. *Right*: The plaque inside Building "D" commemorating
"the master of classical physics." He is remembered for the Eötvös Law, the torsional
pendulum, and the Eötvös Effect.

Left: Loránd Eötvös's bust in the hallway at 1–3 Egyetem Square, District V. *Center*: His grave and bust (by Alajos Stróbl) at Fiume Avenue National Necropolis (plot 10/1-1-9). *Right*: A model of the torsional pendulum in the garden of the former Eötvös Loránd Geophysical Institute, now, part of the Geological and Geophysical Institute of Hungary. The address is 17–23 Kolumbusz (written also as Columbus) Street, District XIV.

József Szabó (1822–1894) first studied philosophy and law, and later mining and mineralogy. He researched the geological characteristics of the Great Hungarian Plane. He founded the mineralogical institute of the University of Budapest

The secessionist style building of the Geological and Geophysical Institute of Hungary at 14 Stefánia Avenue, District XIV. It was built between 1896 and 1899 according to Ödön Lechner's design.

The entrance to the Institute and its top decoration.

and served as professor, dean, and rector. He was active in a multitude of scientific societies and associations. He was a member of HAS. The Geological Institute operates also as one of the departments of Eötvös Loránd University. A former director of the Geological Institute, Miklós Kretzoi (1907–2005) is remembered with a plaque at his former home.

One of the buildings at Trefort Garden is called "Gólyavár," that is, the fortress of storks.[2] There are two tablets on its wall. One commemorates the founding of

SZABÓ JÓZSEF
1822 – 1894
A HAZAI GEOLÓGIAI
TUDOMÁNYOK ÚTTÖRŐJE,
A 125 ÉVES MAGYARHONI
FÖLDTANI TÁRSULAT
ALAPÍTÓ TAGJA EMLÉKÉRE
BUDAPEST FŐVÁROS TANÁCSA
1973

E házban élt és dolgozott
1980-tól 2005-ig
KRETZOI MIKLÓS
(1907 – 2005)
Széchenyi-díjas egyetemi tanár,
paleontológus,
a Földtani Intézet igazgatója.
Születésének 100. évfordulóján
állíttatták
Tisztelői 2007. február 9.

József Szabó's memorial plaque at the corner of Stefánia Avenue and Népstadion Passage.

Plaque honoring Miklós Kretzoi at 24 Lővőház Street, District II.

the Faculty of Philosophy in 1635, and the other notes that in 1809 the botanical garden of the University moved to the location of the present-day Trefort Garden. The "Gólyavár" opened in 1897 and it has a large lecture hall. It first served the Technical University, and since 1910, it has belonged to the University of Budapest. The building was designed by Samu Pecz, professor of architecture at the Technical University. His statue stands at the top of a well, next to a Calvinist church, also designed by him, in District I.

"Gólyavár" (fortress of storks) at Trefort Garden.

Statue of Samu Pecz (by Lajos Berán, 1929), professor of architecture at the Technical University, at the top of a well, on Szilágyi Dezső Square, off Fő Street, District I. Pecz is presented in a master builder's attire.

From among Pecz's designs, we mention here the principal building of the National Archives of Hungary, at 2–4 Bécsi kapu Square, District I. Pecz received the assignment in 1912, but the building opened for operation only in 1923; the delay was due to World War I, the revolutions, and the hard times that followed.

Main building of the National Archives of Hungary at 2–4 Bécsi kapu Square, District I.

There are memorials dedicated to fallen students and instructors at the university campuses and buildings, indicating their sacrifices in the senseless World War I. Two revolutions and a counter-revolution followed, and there were rapid changes in university personnel. In 1919, the short-lived communist dictatorship introduced red terror. However, its university policy brought much, hitherto untapped talent into professorial positions. The future world-renowned aerodynamicist Theodore von Kármán (Chapters 6 and 7) was a powerful officer in the ministry of education at the time. It was ironic that his actions—though indirect, but due to his position—resulted in making some of the best professors in Hungary, eventually, unemployable. When the communists were defeated, the subsequent right-wing regime dismissed all of the professors appointed during the revolutions (not only during the communist dictatorship but also during the preceding bourgeois-democratic regime). George de Hevesy, Béla Bartók, and Zoltán Kodály were among those dismissed, as was the mathematics professor, Manó Beke.

A stone column honors Manó Beke (1862–1946) in the Bibó István Park (which used to be called Barsi Park, and lies between Barsi Street and Bimbó Avenue),

District II. Its hardly visible text says that Manó Beke was a mathematician and professor at the University. He was one of the pioneers of Hungarian mathematics education. What the memorial column for Manó Beke does not say is that he studied in Göttingen and Budapest and started his career as a maths teacher in high school (in Markó Street). From 1896, he was professor of mathematics at the University of Budapest, and between 1911 and 1912 served as Dean of the Faculty of Philosophy. In 1914, he was elected to the Hungarian Academy of Sciences. Then, this brilliant career broke as a consequence of his involvement in university life under the communist regime in 1919. The Hungarian Academy of Sciences excluded him from among its members and the University fired him. Between 1922 and 1939, he served as a mathematical adviser outside academia. In 1945, shortly before he died, his Academy membership was reinstated.

Ruthless "White Terror" followed the suppression of the communist dictatorship. That there were a considerable number of Jewish leaders in the fallen regime was a pretext for violent anti-Semitic actions.

On June 4, 1920, the so-called Trianon Peace Treaty was concluded—named after the Versailles palace where it was signed—with tragic consequences for Hungary. It dismembered historic Hungary by giving independence to Croatia and Slovakia and carved out large chunks of the country to add them to Romania, with some territory also even to Austria. In the "happy peacetime" between 1867 and 1914, the presence of an assimilated Jewish population strengthened the Hungarian population, which was appreciated, because the Hungarian

Memorial to the fallen heroes of World War I (by Béla Horváth, 1936) in Trefort Garden.

population was in a minority in this multiethnic country. After Trianon, there were hardly any sizeable minorities within the new borders.

Many young Hungarians, who suddenly found themselves living outside of Hungary as a consequence of the Trianon Treaty, moved to Budapest seeking higher education. The relatively large Jewish student body became a target and Hungary introduced the first anti-Jewish legislation in post-World War I Europe. The infamous Law XXV in 1920—usually referred to as *numerus clausus*, that is, closed number—severely limited the number of Jewish students at the universities. This happened under Prime Minister Pál Teleki, a noted scientist in geography and a member of HAS (Chapter 5). His reign was short lived, but he would be back as prime minister toward the late 1930s, with more severe anti-Jewish legislation.

The *numerus clausus* remained in effect throughout the Horthy era in spirit, even if its stipulations loosened toward the late 1920s. Kuno Klebelsberg (1875–1932), the minister of religion and public education between 1922 and 1931 and a member of HAS, proposed replacing references to racial and national origin in the law by national loyalty and moral reliability. Teleki, Klebelsberg, and their colleagues understood that the *numerus clausus* legislation reflected negatively on Hungary in the civilized world, and Klebelsberg practised a sophisticated double talk. In the West he said what he supposed was acceptable there and at home he continued the rhetoric that served domestic politics. In 1924, he made this cynical statement in the Hungarian Parliament, as though shifting the responsibility to the West: "Give us back the old Greater Hungary and then we will be able to revoke numerus clausus."[3]

As for women students, in 1926, Klebelsberg terminated their 1919 exclusion from university studies at the Faculty of Medicine (the exclusion continued until 1945 at the Faculty of Law and State Affairs).

Klebelsberg had a broad vision for the dominance of Hungary in the region through Hungarian "cultural superiority." He aimed at bringing back some of the talent that had left the country. It is doubtful though that he would have welcomed the return of Jewish expatriate scientists. None of the attempts to gain professorial appointments for Jewish physicists or mathematicians turned out to be successful, although they had in the meantime become top players, internationally, in their fields.

Max Born, the doyen of the famous Göttingen school of physics tells an amusing story. It happened in about 1930 that Klebelsberg visited this famous university and its chief administrator, J. T. Valentiner,[4] gave a luncheon party for the distinguished visitor to allow him to meet with the luminaries of the school. This is how Born remembered it when the high guest asked him what he [Born] "thought about the Hungarian mathematicians and physicists. I replied with a hymn of praise for my Hungarian colleagues, mentioning first my old friends [Alfred] Haar and [Theodore von] Kármán, then [George] Polya in Zürich and others I cannot now remember, and finally the young generation who were at pre-

Above: Statue of Kuno Klebelsberg, at Villányi Avenue, District XI, between Szent Imre Gimnázium of the Cistercian Order and Szent Imre Church. The female side figure symbolizes the arts and the male symbolizes science. The work of sculptor Jenő Grantner was unveiled in 1939 elsewhere in Budapest. The statue suffered damage in the war and it was removed after the war. Sculptor Kálmán Tóth reconstructed it, and since 2000 it has stood in its present location. *Right*: Kuno Klebelsberg's bust in the yard of Puskás Tivadar Technical High School of Telecommunication, 22 Gyáli Avenue, District IX.

GRÓF
KLEBELSBERG
KUNO
1875 – 1932

sent in Göttingen: John von Neumann, Eugene Wigner and Edward Teller. At this point, I got a fearful kick on the shin from [James] Franck, whereupon I stopped and let him continue the discussion. I did not understand what he meant by this violent interruption until he explained it to me after lunch. All I had mentioned were Jews, and therefore, in the eyes of a representative of that anti-Semitic government, not Hungarians at all."[5]

In the late 1930s and early 1940s, the *numerus clausus* gradually became *numerus nullus*, and there was increasing discrimination against the remaining Jewish students. As war was approaching, para-military instructions became more rigorous. From 1942, Jewish students had to wear yellow armbands and the converted Jewish students white armbands. Thus, such visible anti-Jewish discrimination predated the German occupation of Hungary (on March 19, 1944) after which the Jews in Hungary had to wear the yellow star. In the early summer of 1944, many Jewish university instructors and students were deported to concentration camps, where most of them perished.

After the liberation of Hungary in 1945, for a few years, the University continued its operations, still consisting of four faculties (theology, medicine, law, and philosophy). In 1949, an important reform was the division of the Faculty of Philosophy into two, the Faculty of Humanities and the Faculty of Science (natural sciences and mathematics). From the academic year 1948/49, for the next four decades, the authorities thoroughly politicized university life. University autonomy disappeared, and the communist party exercised full control over all aspects of university operations.

From 1921, the University of Budapest carried Péter Pázmány's name as the Budapest Royal Hungarian Pázmány Péter University. In 1950, the Faculty of Theology was abolished, and in the same year, the University received a new name. It proved to be an excellent choice, Loránd Eötvös, but the authorities did not even consult with the University about it. In 1951, again without involving the University, the authorities separated the Faculty of Medicine from the rest of the school and it continued as Budapest University of Medicine. The Faculty of Humanities was moved from Trefort Garden to the magnificent buildings of the Gimnázium of the Piarist Order and the Gimnázium was forced out for the next half century to another location. However, the Piarists fared better than some other denominational high schools which had to cease their operations altogether.

The original founder of the university, Péter Pázmány, remained without an institution of higher education for four decades. In the 1990s, after the political changes, there was then a rebirth of a Pázmány Péter Catholic University, whose origin dates back 1635. It has several faculties, mostly located in Budapest. The Facuty of Law and Political Sciences is at 28–30 Szentkirályi Street, District VIII.

Pázmány Péter Catholic University—Faculty of Law and Political Sciences—with St. István's statue, at 28–30 Szentkirályi Street, District VIII.

The main building of the Central European University at 9 Nádor Street, District V.

There have been other new developments in the network of institutions of higher education; among them, other denominational schools have appeared. There is also a non-denominational private school. The Hungarian-born American financier George Soros founded the Central European University (CEU) in 1991 and his generous endowment has supported it ever since. It is an institution for graduate studies in social sciences, the humanities, law, business, and management.

———— • ◆ • ————

A bust pays tribute to István Bibó (1911–1979), lawyer, civil servant, political theorist, and former associate of the Library of Eötvös Loránd University. His most productive period was in the short-lived democratic period from 1945 to 1948, when he wrote about the misery of the small states in Eastern Europe, the crisis of Hungarian democracy, and the Jewish question in Hungary after 1944. His works had great impact among twentieth-century Hungarian thinkers. On November 3, 1956, Bibó was appointed Minister of State in Imre Nagy's government. The Soviet invasion came the next day, and Bibó stayed the longest among the government members in the Parliament, until November 6. Later he was arrested, tried, and sentenced to life imprisonment, but was released in 1963.

Left: István Bibó's bust on Széchenyi Quay (by Géza Széri-Varga, 2005); *right*: statue of György Lukács in St. István Park, District XIII (by Imre Varga, 1985).

There is a György Lukács (1885–1971) statue in St. István Park. Lukács was a Marxist philosopher of international fame, influential literary critic, and communist politician. He was born into a wealthy ennobled Jewish–Hungarian family. He started publishing philosophical works from his early youth, and participated in actual political movements. He had high-level positions in the 1919 communist dictatorship as deputy people's commissar for education and then as the political commissar of the Hungarian Red Army. After a military defeat against Romanian invaders, he committed an infamous act by having eight Red Army soldiers executed by decimation, that is, by selecting the victims as every tenth man in a line-up.

Between the two world wars, Lukács lived in exile, first in Western Europe, then from 1930 in the Soviet Union, which he could not leave until 1945. After the war, he supported the communist dictatorship in Hungary, and contributed to the persecution of intellectuals. He himself was accused from time to time of rightist deviation from the true Marxist doctrine. In 1956, he was a member of Imre Nagy's short-lived revolutionary government, but suffered few consequences thereafter. He built up an influential school of followers of his philosophy, in both Western Europe and Hungary. He aimed to reconcile a failed political system with an appealing ideology—which proved to be an impossible task. He taught aesthetics at Eötvös University and was a member of HAS. There is a memorial plaque on the house where Lukács lived, at 2 Belgrád Quay, District V.

There is a memorial plaque honoring Baruch Spinoza—Benedict de Spinoza (1632–1677)—in Dob Street. He was a most influential Jewish–Dutch philosopher, a forerunner of the eighteenth-century Enlightenment and a critic of the Bible.

———— • ◆ • ————

Baruch Benetictus Spinoza's memorial plaque (by Antal Czinder, 2004) on the wall of "Spinoza House," at 15 Dob Street, District VII. The rough translation of the Spinoza quote on the plaque: Weapons can't conquer the spirit; only love and generosity can. The world-renowned scientist, author, and playwright Carl Djerassi (*on the left*) visited the memorial in May 2013.

After many years of planning to build a new campus in the spacious Lágymányos in District XI, the first building, for the chemistry departments, opened in 1989. Physics and biology followed, and the humanities gradually occupied the old, renovated campus in Trefort Garden. We have previously mentioned some memorabilia of scientists that were left in Buildings "B" and "D". Other memorabilia

There is a bust (by Bernadett Szilágyi) of Károly Than (1834–1908) in the spherical aula. He was the founder of scientific chemistry in Hungary. His tombstone is in the Fiume Avenue National Necropolis (plot Á. B. 9). A plaque honors him in the chemistry department.

Béla Lengyel (1844–1913), professor of chemistry. His relief is in the hall of the chemistry department and the bust (by Mária D. Törley) is in the spherical aula.

Lajos Winkler (1863–1939), professor of chemistry. His relief is in the hall of the chemistry department and the bust (by Aranka E. Lakatos) is in the spherical aula.

Gusztáv Buchböck (1869–1935), professor of chemistry. His relief is in the hall of the chemistry department and the bust (by Béla Tóth) is in the spherical aula.

were moved to the new campus. There is also a set of busts of great Hungarian scientists in its impressive spherical aula.

Lengyel, Winkler, and Buchböck were members of HAS, and they, along with other professors, founded the "Kis Akadémia"—the Little Academy—in 1899. It was an informal association of mostly natural scientists. Its goal was fostering friendship among its members and the dissemination of scientific information. The association published a prestigious book series containing the memorable presentations of its members.

There are further memorial plaques, not shown here, in the hall of the chemistry department. Gyula Gróh (1886–1952) studied in Budapest and in Germany. He initiated the special training of chemists in Hungary. Together with George de Hevesy, he participated in the introduction of radioactive isotopes as tracers in scientific research. He was professor of chemistry at the veterinary school (see Chapter 5), then at the Budapest Technical University (Chapter 6), and, finally, at the University of Budapest. In 1950, for unknown reasons, he was forced to retire from teaching, but he continued research at a food industry laboratory. He was a member of HAS, but in 1949, the Academy downgraded his membership to the status of "adviser." In 1989, his membership was reinstated, posthumously.

Győző Bruckner (1900–1980) was professor of organic chemistry first at Szeged University, later in Budapest. He was an internationally recognized peptide chemist with important achievements in development of pharmaceuticals. He authored a comprehensive set of organic chemistry textbooks. He was a member of HAS.

László Szebellédy (elsewhere, Szebellédi, 1901–1944) was a professor of chemistry whose main research interest was in inorganic chemistry and analytical chemistry.

Tibor Erdey-Grúz (1902–1976) was professor of physical chemistry, his main interest being electrochemistry. He wrote textbooks and popularized science for lay readers. He was also a politician and between 1952 and 1956, he was a member of the government, first as minister of higher education and then as minister of education. Over the period 1954–1956, he was a member of the central committee of the communist party. He was a member of HAS and occupied leading positions in the Academy.

Elemér Schulek (1893–1964) has a plaque in the hall of the chemistry department and another on the building where he lived, 10–12 Bartók Béla Avenue, District XI. He was a pharmacist and analytical chemist focusing on the analysis of pharmaceuticals. He was professor of analytical chemistry and a member of HAS.

Plaques of Elemér Schulek and Mózes Rubinyi at 10–12 Bartók Béla Avenue.

There is yet another plaque at 10–12 Bartók Béla Avenue, honoring Mózes Rubinyi (1881–1965), historian of literature. Most of his life, Rubinyi taught in secondary schools, with the exception of one brief period in 1919 during the communist dictatorship and again between 1947 and 1951, when he held a professorial appointment. In 1948, the HAS elected him as a member, but in 1949, it downgraded his membership. In 1989, his membership was reinstated posthumously.

Aladár Buzágh's relief in the hall of the chemistry department and his tombstone (joint with his wife) in Farkasrét Cemetery.

Aladár Buzágh (1895–1962) was a chemical engineer and earned a doctorate at the University of Budapest. He founded research in colloidal chemistry in Hungary. He was professor of chemistry and a member of HAS.

We have already met some of the busts that stand in the spherical aula on the new campus located in the physics building. Students like to sit on the stone benches among the statues to study, chat, or just relax. Here we present the remaining busts.

Bust of Károly Tangl (by Frigyes Janzer) in the spherical aula.

Bust of Rudolf Ortvay (by Bernadett Szilágyi) in the spherical aula.

From left to right, Rudolf Ortvay, Paul Dirac, and Eugene P. Wigner in the early 1930s in Hungary. Dirac was already a Nobel laureate and married Wigner's sister.
Courtesy of the late Martha Wigner Upton.

Károly Tangl (1869–1940) was a physicist, a member of HAS, and professor of the University of Budapest and the Budapest Technical University. He did research in a variety of areas of experimental physics, including cosmic rays, magnetism, the dielectric properties of gases and liquids, and capillarity.

Rudolf Ortvay (1885–1945) was a physicist member of HAS. He studied in Budapest, Göttingen, Zurich, and Munich. He taught at Szeged University and at the University of Budapest. He founded a school of theoretical physics, corresponded with the great emigré scientists, and helped maintain an international outlook for the physics community in Hungary. In 1945, he committed suicide for unknown reasons.

Radó Kövesligethy (1862–1934) was an astronomer and physicist and a member of HAS. He studied in Vienna and at one time, he was Loránd Eötvös's assistant. He was professor of physics at the University of Budapest. In the early 1900s, he founded the Budapest Seismological Observatory. He held a position during the 1919 communist rule and was not allowed to teach for years afterwards. Astronomy and earthquake research were his main areas of activity and he contributed much to the establishment of the national monitoring system to forecast earthquakes.

A closely related memorial plaque, but not at the university, honors Miklós Konkoly-Thege (1842–1916). He was an astronomer and meteorologist and a member of HAS. In 1871, he initiated a private observatory in the park of his

Radó Kövesligethy's bust (by Aranka E. Lakatos) in the spherical aula. The plaque is on the building of the Kövesligethy Radó Seismological Observatory at 18 Meredek Street, District XI. The plaque also honors László Egyed (1914–1970) who in the mid-1950s reorganized the seismological network and earthquake research.

castle in Ógyalla (then, in northern Hungary; today, in Slovakia). In 1899, he donated this observatory to the Nation. He organized the Hungarian weather forecasting service by means of a telegraph network. The memorial plaque at 6 Fő Street, District I, states that between 1892 and 1910, the Royal Hungarian Central Institute for Meteorology and Earth Magnetism operated on two floors of this building under Director Miklós Konkoly-Thege.

The plaque honors Miklós Konkoly-Thege at 6 Fő Street, District I. The statue (by Ferenc Sidló) is the tombstone on his grave at the Fiume Avenue National Necropolis (plot 37/2-island).

József Száva-Kováts (by Frigyes Janzer), meteorologist.

László Detre (by István Szabó, Jr.), professor of astronomy in the spherical aula of Eötvös University.

The meteorologist József Száva-Kováts (1898–1980) founded the first department of atmospheric and climate science in Hungary and organized the training of meteorologists at Eötvös University. László Detre (1906–1974), astronomer, received his training in Hungary, Germany, and Austria. He was professor of astronomy.

Georg von Békésy, György Békésy in Hungarian, was a Nobel laureate in Physiology or Medicine (see Chapter 1). He spent much of his scientific career at the research laboratory of the Hungarian Post Office in Budapest. Between 1939 and 1946, he was professor of experimental physics at the University of Budapest. In 1946, he left for Sweden, and in 1947, he moved to the United States. He was at Harvard University between 1947 and 1966 and at the University of Hawaii in Honolulu until 1972 (for further details of his career, see Chapter 6).

George de Hevesy was a Nobel laureate in Chemistry (1943, see Chapter 1). He was born in Budapest as György Bischitz. The name "hevesy" was added in front of his father's name on the occasion of his ennoblement as though indicating that the family estate was in Heves. Soon the family switched from "hevesy Bischitz" to Hevesy. Hevesy signed his German

George von Békésy's bust (by György Benedek, 1999) in the spherical aula.

publications as Georg von Hevesy and he inserted "von" to make sure his name conveyed that he was a member of the nobility. After the Nazi takeover of Germany, Hevesy signed his papers as George de Hevesy.

Hevesy had Jewish roots on both sides, belonging to Hungarian nobility on both his mother's and his father's side. It was not rare in the last decades of the nineteenth century for Franz Joseph I to ennoble Hungarian families of Jewish origin, but the ennoblement of Hevesy's maternal side was one of those rare cases when the King accorded this honor to unconverted Jews.

Hevesy had received a Catholic upbringing and studied at the Gimnázium of the Piarist Order in Budapest. He started his higher education at Budapest University and continued at the Technical University of Berlin and the University of Freiburg, Germany, earning his doctorate in chemistry. He did research at a number of leading European laboratories in Austria, Britain, Denmark, Germany, Switzerland, and elsewhere, and did joint work with several of the scientific giants of the twentieth century, including Ernest Rutherford and Niels Bohr.

From time to time, he renewed his ties with Hungary, and in 1915, he voluntarily joined the Austro-Hungarian Army. His peripatetic nature resulted from always seeking the most appropriate conditions for his research and also because he could not find adequate employment at home. This changed in 1918 when Loránd Eötvös initiated Hevesy's professorial appointment at the University of Budapest during the last months of the Austro-Hungarian Monarchy. Still in 1918, but already under the Republic of Hungary, the University forwarded the proposal to the ministry of education. Finally, his appointment materialized on May 3, 1919, when Hungary was already under communist dictatorship.

Hevesy dedicated himself to his university activities with zest and von Kármán in the ministry of education gave full backing for Hevesy's innovations. When the White Terror took over, von Kármán had to flee the country, and the university authorities accused Hevesy of cooperating with von Kármán and the red dictatorship. Hevesy was humiliated and he lost his right to lecture to students. He decided to move back to Western Europe where all of the doors opened to him that were closed in his homeland.

George de Hevesy's bust (by Béla Tóth) is in the spherical aula.

Hevesy summarized his plight in a letter, dated October 25, 1919, to his friend, Niels Bohr: "Politics entered university as well. In my absence, during the summer they fired two honest and intelligent assistants of mine, just because they were Jewish, and as far as I know this happened in the other institutes, too. Jews and suspected radicals could not keep their jobs. Under such circumstances, I could do nothing, but resign from my position."[6] He told Bohr about von Kármán: "The university authorities hate him for he held the position of being in charge of higher education under the communists, although von Kármán was not a communist. He was the one who prevented the communists from destroying the universities. The current destruction both in a material and in a moral sense will prevent a successful university and scientific life in Hungary."[7] At this time, Hevesy left Hungary for good.

He spent a happy period in pre-Nazi Germany. He was a Hungarian aristocrat and nobody knew about his Jewish roots. When the Nazis came to power, he did not wait for the racial laws to reach him; in 1934, he resigned his position as though in protest, and continued his life and work in Copenhagen. Although in 1940 the Germans occupied Denmark, he stayed until 1943, when he left for Sweden. There he continued for almost two decades but for his last years he returned to Freiburg, Germany.

Apart from brief visits, his final return to Hungary would only come in 2001, when the Hungarian Academy of Sciences reburied his ashes at the academic section of the Fiume Avenue National Necropolis.

Graves of scientists at the academic section of the Fiume Avenue National Necropolis: On the right, George de Hevesy's grave is the second from the bottom. Leo Szilard's grave is on the left, second from the bottom (a portion of Szilard's ashes are buried in Ithaca, NY). The inscriptions on these uniform graves are simple: In addition to name and years of birth and death, membership of HAS is mentioned. Only Szilard's inscription is different—he was not a member of HAS.

Hevesy received many nominations for the Nobel Prize in Chemistry, and in 1944, for 1943, the Royal Swedish Academy of Sciences finally gave it to him. In the course of his nominations, the nominators mentioned a variety of his discoveries. The prize announcement singled out one of them, "for his work on the use of isotopes as tracers in the study of chemical processes." Various anecdotes have described the original discovery and even if it is impossible to vouch for their authenticity, they help us to grasp its essence.

Here is one of them. Hevesy was staying in a boarding house, and noticed a similarity between leftover meat pieces from the previous day's dinner and the meat pieces in the soup the following day. He suspected that the kitchen recycled leftover beef. He wanted to find a way to confirm his suspicion. One evening he left a piece of meat on his plate and smeared on it a little substance containing a radioactive isotope. The amount was too small to harm anyone, but sufficient to detect with a Geiger–Müller counter. Hevesy tested the soup the next day, and his counter showed radioactivity. Supposedly, this was the first practical use of radioactive tracers.

Here we mention a few mathematicians of exceptional fame and achievement, alas, there are hardly any remaining visible commemorations related to them apart from their tombstones. Lipót (Leopold) Fejér (1880–1959) was a world-renowned mathematician and mentor to generations of mathematicians. He produced seminal results in the area of Fourier series, interpolation theory, functional analysis, among others. He was professor of mathematics at the University of Budapest and a member of HAS.

Frigyes Riesz (1880–1956) was co-creator—with Alfred Haar—of the famous school of mathematicians at Szeged University. His main achievements were in functional analysis, set theory, topology, projective geometry, among others. He was professor of mathematics at the University of Budapest and a member of HAS.

Left: Memorial plaque on the building where Lipót Fejér lived, 20 Bajza Street, District VI. *Right*: Tombstone of his grave (plot 34/2-1-5) at the Fiume Avenue National Necropolis.

At the unveiling ceremony of Lipót Fejér's plaque at 12 Fejér Lipót Street, District XI. *From left to right*: László Fejes Tóth, unknown, Vera T. Sós, Miklós Hosszú, György Alexits, István Vincze, and Pál Turán.
Courtesy of Vera T. Sós, Budapest.

Cornelius (Kornél) Lánczos (1893–1974), one of Fejér's pupils, was a world-renowned Hungarian–American–Irish mathematician/theoretical physicist. His main interest was in the unification of geometrical theories of gravitation and electromagnetism. He studied in Budapest and Szeged, and beside Fejér, his mentors included Loránd Eötvös and Rudolf Ortvay. In 1920, to escape anti-Semitism, he left Hungary and spent the rest of his life in the United States and in Ireland. Between 1928 and 1929, he served as Albert Einstein's assistant. He was a member of the Irish Academy of Sciences. Most of his family members perished in the Hungarian Holocaust. Lánczos died during a visit to Budapest.

Pál Turán (1910–1976) was a member of HAS who gained international renown in number theory, analysis, theory of graphs, statistics, theory of probability, among others. He studied at Budapest University, graduated as a mathematics teacher in 1933, and acquired his doctorate in 1935. For years he was

The grave of Frigyes Riesz (plot 34/2-1-6) at the Fiume Avenue National Necropolis.

The grave of Kornél (Cornelius) Lanczos (plot B-3-3) in the Jewish section of Farkasréti Cemetery.

Photograph courtesy of József Varga, Budapest.

unemployed, until he was appointed to teach mathematics in the high school of the Budapest School of Rabbis. From 1940, he served as a slave laborer for extended periods of time, as prescribed by anti-Jewish legislation. In spite of harsh conditions, he never stopped his research activities in mathematics. His career could take off only after World War II. He was chair and professor of algebra and number theory at Budapest University and had visiting professorships in numerous institutions abroad.

Alfréd Rényi (1921–1970) was a member of HAS and directed the Mathematical Research Institute of the Hungarian Academy of Sciences—today, Rényi Institute. He contributed significantly to many areas of mathematics, initiated the Hungarian school of probability calculations and his results have been utilized not only in mathematics, but also in physics (quantum mechanics). He was interested in the philosophical questions of mathematics, started research into the history of ancient mathematics, and started reforming mathematical instructions in high schools.

Pál Erdős (1913–1996)—known better as Paul Erdos—was one of the most famous mathematicians of the twentieth century; he was also one of the most colorful and most prolific of scientists. He was a member of a friendly circle of young mathematicians in Budapest in pre-World War II times whose lives revolved around their beloved subject, but who faced persecution to an ever increasing degree. Erdős started spending time abroad and from 1938

The grave of Pál Turán (plot C1-12-1) in the Jewish section of the Farkasréti Cemetery. The names of Turán's three siblings murdered in the Holocaust are also listed on the tombstone.

Photograph courtesy of József Varga, Budapest.

The grave of Alfréd Rényi (plot 6/1-1-31) at Farkasréti Cemetery.

Photograph courtesy of József Varga, Budapest.

Plaque on the wall of the house where Alfréd Rényi and his family lived at the corner of Munkácsy Mihály and Délibáb streets in Distric VI. Rényi's wife, herself a mathematician, is also mentioned on the plaque.

he stayed away from Hungary for almost two decades. Even though he visited Hungary frequently from the mid-1950s, he spent much of his time visiting international colleagues. His only permanent affiliation was with the Rényi Institute of the HAS in Budapest, which gave him complete freedom of movement.

Erdős was peripatetic and had about five hundred co-authors on his papers. He contributed to various diverse areas of mathematics; he solved problems,

Paul Erdős with his beloved mother in 1969 being drawn by Kathy Alpar-Altman.
Photograph courtesy of Vera T. Sós, Budapest.

Erdős and his mother lived in this apartment house, 8 Abonyi Street, District XIV.

and recognized problems for whose solution he offered rewards—often very small ones—but winning even a small reward from Erdős counted as great recognition. He gained a benevolent notoriety and mathematicians and others have determined how close they got to him by establishing the so-called Erdős Number. His own number was 0. Any co-author of his collaborative papers had Erdős Number 1. If one co-authored a paper with someone who had Erdős Number 1, that person had Erdős Number 2, and so on. Erdős had an impact on the development of current mathematics and on recognizing the importance of collaborative effort.

Grave of the Erdős family at the Kozma Street Jewish Cemetery (plot 17A-6-29). Erdős's and his mother's names are carved into the smaller stone. The main stone carries the names of Erdős's two sisters (who died of illnesses at the time of his birth in 1913), and the name of his father.

The scientific visitor may be surprised to see an unusual tombstone in the shape of a pentagon dodecahedron, one of the five Platonic solids, near the graves of Lipót Fejér and Frigyes Riesz at the Fiume Avenue National Necropolis. One might expect it to belong to a mathematician, but this is not the case; it decorates the writer Béla Illés's (1895–1974) grave. Illés earned his law degree at the University of Budapest and Endre Ady discovered his talent in writing. Illés became involved in socialist politics early in his life and had to emigrate after 1919. From 1923, he lived in Moscow. In 1945, he returned to Hungary as an officer of the Red Army. He was a natural storyteller.

Vladimir N. Gribov's (1930–1997) grave is another unusual one. He was a Russian theoretical physicist, first in Leningrad, then, in Moscow. He attended Lev Landau's famous seminars and later Gribov developed his own seminars on quantum field theory and elementary particles. He was a senior scientist at

Left: Béla Illés's grave in the shape of a pentagonal dodecahedron (plot 34/2-1-24). *Right*: Vladimir N. Gribov's grave (by Imre Varga, 1998, plot 42/2). Both are at the Fiume Avenue National Necropolis.

the Landau Institute in Moscow and during his last years at the Central Physics Research Institute in Budapest. His tombstone, "Eternal Matter," symbolizes the atom. A transparent blue stone in the middle represents the nucleus and the vaulted chrome steel around it represents the electron domains.[8]

Notes

1 László Szögi, ed., *Az Eötvös Loránd Tudományegyetem története 1635–2002* (Budapest: ELTE Eötvös Kiadó, 2003).

2 The origin of this name is not clear; it may refer to the birds or it may refer to the freshman students who in university language are known as "storks."

3 Kovács M. Mária, *Törvénytől sújtva: A numerus clausus Magyarországon, 1920-1945* (Budapest: Napvilág Kiadó 2012), p. 50 idézi: *Nemzetgyűlési Napló*, 1922–1926. XXIV. Kötet, 295. ülés, 320. old. (1924. június 4.)

4 Just an aside, Valentiner would soon dismiss the famous female mathematician Emmy Noether from university employment. He declared that "Noether was too much of a socialist and a left-winger to ever make a good Nazi." However, the official paper merely cited her Jewish background. Hitler wrote in his *Mein Kampf*, "The message of women's emancipation is a message discovered solely by the Jewish intellect, and its content is

stamped by the same spirit." Joseph Goebbels added, "The mission of women is to be beautiful and to bring children into the world." [*Nobel Prize Women in Science* by Sharon Bertsch McGrayne, p. 83.]

5 Max Born, *My Life: Recollections of a Nobel Laureate* (New York: Charles Scribner's Sons, 1978), p. 236.

6 Gábor Palló, *Hevesy György* (Budapest: Akadémiai Kiadó,1998), pp. 119–120.

7 Gábor Palló, *Hevesy György*, pp. 120–121.

8 Private communication from Júlia Nyíri (Vladimir Gribov's widow), March 6, 2014.

4
Ignác Semmelweis
MEDICINE

Ignác Semmelweis. His bust stands in the inner campus of Semmelweis University. Semmelweis University has two principal campuses and additional locations for its institutions. Both campuses are along Üllői Avenue, in District VIII. The inner campus is bordered by Üllői Avenue-Mária Street-Baross Street-Szentkirályi Street. The outer campus is bordered by Üllői Avenue-Korányi Sándor Street-Illés Street-Tömő Street-Balassa Street-Apáthy István Street-Szigony Street.

Ignác F. (Ignaz Philipp) Semmelweis (1818–1865) was born in Buda into a well-to-do German speaking Catholic family, which considered itself to be Hungarian.[1] Today, the building where he was born houses the Semmelweis Museum of the History of Medicine. Semmelweis started his schooling at a time when the language of instruction was still Latin. He graduated from the Royal University Catholic Gimnázium in Buda, the oldest high school in Hungary (which no longer exists). Some of its teachers conducted independent research, and József Eötvös was among its former pupils.

In 1837, Semmelweis began his studies as a law student in Vienna, but soon transferred to the medical school, from which he graduated in 1844. He did part of his studies in Pest. From the start, he was interested in pathology. In 1846, he was appointed assistant professor at the 1st Department of Obstetrics of Vienna University. The frequent incidence of puerperal fever—childbed fever—leading to death appalled Semmelweis, and he set out to change the situation. His painstaking studies led to important observations.

He noticed that the frequency of puerperal fever and maternal mortality rate were much higher at the 1st department than at the 2nd. The 1st department trained doctors and the 2nd department trained midwives. The more modern approach to obstetrics of the 1st department prescribed autopsy of cadavers, whereas such autopsies were rare at the 2nd department. Further, Semmelweis observed that the numbers of cases of puerperal fever and deaths was very low among women who gave birth at home rather than going to a clinic. Well-to-do women could more easily afford to stay at home for childbirth than could poorer women. However, social differences did not provide a straightforward explanation for the difference in mortality rates. Then there was the tragedy of a professor who injured himself during an autopsy and died of very much the same symptoms as the women in the obstetrical clinics. Finally, Semmelweis's

Left: The first Semmelweis statue (by Alajos Stróbl, 1906) in front of the St. Roch Hospital, at the corner of Rákóczi Avenue and Gyulai Pál Street, District VIII. *Right*: Semmelweis statue (by Béla Domonkos, 2004) in the inner campus of Semmelweis University.

most disturbing observation was that when he spent a few months away from the clinic, the mortality rate dropped noticeably. When he was present, he performed a particularly large number of autopsies.

Semmelweis's milestone discovery was that the obstetricians themselves and their students transported the poisoning ingredients from the autopsy to the women whom they examined in the clinic. This was a revolutionary realization at the time which preceded subsequent understanding of the germ theory of disease and the existence of bacteria as carriers of disease. This added to the difficulties Semmelweis faced in gaining acceptance for his suggestions. He proposed the introduction of disinfection before medical personnel touched a patient, after they had performed a postmortem examination. Rather than the usual water and soap for washing hands, he prescribed the use of a solution of chlorinated lime (corresponding with the common chlorine bleach solution in households today) to destroy the poisonous cadaveric agents. His approach led to spectacular results in diminishing the number of women dying in childbirth. His achievements have been compared with Edward Jenner's discovery of vaccination against smallpox one generation before. Jenner though, in England, had a smoother life and fast acceptance of his teachings.

One would have expected that after Semmelweis's success in discovering the reason for puerperal fever and childbed death and then providing the remedy that drastically changed the situation, he would have had a straightforward path to promotion and glory. Alas, this was not at all the case. He had difficulties in Vienna, and when he left and continued his career in Pest, the difficulties went with him. It was not so much the lack of recognition that bothered him, but the fact that in his Viennese clinic, the old practices that caused so many unnecessary deaths continued for many years, unchanged.

He was not happy in his private life either; his much younger wife had no idea about the importance of his work. He was becoming increasingly irrational, but the response of those around him was out of proportion. They locked him up in a lunatic asylum in Vienna where his "nurses" beat him up and left his injuries untreated—it was tantamount to killing him. This happened in 1865; he was 47 years old. The symptoms of his death were similar to those of puerperal fever since the cause was similar too; blood poisoning was a consequence of his untreated wounds. In his death, he provided yet another testimony to his seminal realization.

During his lifetime, Semmelweis never received any award or prize and the Hungarian Academy of Sciences did not elect him to its membership. The recognition he received was the women whom he saved. It took decades following his death for the medical profession and broader society to understand his greatness. Some called him the founder of medicinal antisepsis—disinfection—and the founder of protection from infection—medicinal asepsis. Statues and plaques have since been erected honoring him. UNESCO declared 1965 to be Semmelweis Year. In 1969, the Budapest University Medical School was named after him, Semmelweis University of Medicine, and in 2000, Semmelweis University.

"Semmelweis az anyák megmentője"— Semmelweis the Savior of Mothers and the street sign above the relief, at 1 Semmelweis Street, District V.

It was the ordinary people who accorded him his greatest honor: His name is forever associated with a label, "Ignác Semmelweis—Savior of Mothers."

There is a tablet at the entrance, which informs visitors about Semmelweis's birthplace and resting place of his ashes which are buried in the wall of the courtyard of the Semmelweis Museum. In front of this wall, there is a statue of a mother holding her child. Ignác Semmelweis's journey was not an easy one, but he has finally arrived home.

There is yet another Semmelweis statue in the inner campus, on the façade of the Department of Obstetrics and Gynecology.

Ede Flórián Birly (1787–1854) was Semmelweis's predecessor as professor of obstetrics at the University of Pest. Childbed death was rare in Birly's clinic

Memorial monument for Ignác Semmelweis at the Fiume Avenue National Necropolis without his remains (plot 34/2-1-1).

Ignác Semmelweis's birthplace—today, the Semmelweis Museum of the History of Medicine, at 1–3 Apród Street, District I. As of preparing this book (2014), the building is being renovated and will soon open in renewed splendor.

as compared with Vienna, and Birly ascribed this to his advocacy of purgatives. Birly believed that the uncleanliness of the bowel caused the childbed fever. When, in Semmelweis's words, "medicine in Pest became anatomically oriented,"[2] meaning that the obstetricians began the practice of autopsy, the purgatives no longer prevented childbed fever. Alas, Birly never accepted Semmelweis's discovery.

There is a similar building housing the Department of Transplantation with two statues on its façade.

Semmelweis's burial place in the Semmelweis Museum, with a statue of a mother holding her child.

Façade of the central part of the Department of Obstetrics and Gynecology at 27 Baross Street, District VIII, and its two statues: *on the left*, Ede Flórián Birly and *on the right*, Ignác Semmelweis.

Façade of the central part of the Department of Transplantation and Surgery at 23 Baross Street, District VIII, and its two statues: *on the left*, Sándor Lumnitzer and *on the right*, János Balassa.

Sándor Lumnitzer (often, Lumniczer, 1821–1892) was a surgeon and professor of medicine at the University of Pest. He was the great-grandfather of János Szent-ágothai (see, below). Lumnitzer studied in Pest and in Vienna, and János Balassa was one of his mentors. Lumnitzer participated in the War for Independence, 1848–1849, and was punished for it. His university career took off only after the Compromise. He was professor of medicine and director of the 2nd Department of Surgery from 1880 until his death. During his directorship, he initiated the antiseptic treatment of wounds in Hungary.

János Balassa (1814–1868) was a surgeon; he studied in Pest and Vienna. For years he worked in Vienna, but in 1843 he received his appointment to the University in Pest. He was involved with the War for Independence, 1848–1849, and for a while he was incarcerated. In 1851, he returned to his professorial job and he continued his active involvement in university matters to the end of his life. He created modern surgical training and practice in Hungary.

Balassa initiated a plethora of new surgical procedures in a variety of areas of human medicine. In 1847, he was the first in Europe to apply general anesthetics, using ether narcosis, and was among the pioneers of laryngotomy. Together with Lajos Markusovszky, Balassa founded the *Orvosi Hetilap* (Medical Weekly)—a publication that has remained popular ever since. He was a member of HAS. On the outer campus, there is memorial plaque to János Balassa at the

János Balassa's grave by József Engel, Fiume Avenue National Necropolis (plot J. 194/195).

Pál Bugát's tombstone (by László Csontos), Fiume Avenue National Necropolis (plot J. 169).

entrance to the Neurological Clinic at 6 Balassa Street, District VIII. He is buried at the Fiume Avenue National Necropolis.

Pál Bugát (1793–1865) has a conspicuous tombstone at the Fiume Avenue National Necropolis. He studied in Pest, acquiring his physician's diploma and his certificate of ophthalmologist specialization. Eventually, he was appointed professor and taught physiology, pathology, and pharmacology. Following the suppression of the War for Independence, the authorities took his chair from him and deprived him of his right to a pension. He withdrew from medicine and devoted his remaining years to linguistics. He did considerable work on science popularization also.

The Semmelweis Museum is rich in medical memorabilia. Its busts include one of János Balassa and four others shown below.

Ignác Hirschler (1823–1891) was an ophthalmologist, a graduate of Vienna University, and worked as an instructor at the University of Paris before returning to Hungary. His career exemplified the changing opportunities for Jewish

Busts from the collection of memorabilia at the Semmelweis Museum of Medicinal History; (*top left*) Ignác Hirschler; (*top right*) Vilmos Tauffer; and (*bottom left*) István Tóth. *Bottom right*: The bust of József Béres is in the museum stairway.

professionals before and after the Compromise in 1867. In Pest, he became a sought-after ophthalmologist, but when in 1851 he applied for the title of private docent at the University of Pest, they rejected him because he was Jewish. Later, his career took off; he was elected as a member of HAS and was appointed to membership of the Upper House of Parliament. János Arany was among his patients.

Vilmos Tauffer (1851–1934) was professor and director of the 2nd Department of Gynecology and Obstetrics. He faithfully followed Semmelweis's teachings and created a school of Semmelweis's adherents. István Tóth (1865–1935) was Tauffer's

Right: The plaque of József Antall in the entrance hall of the Semmelweis Museum. *Below*: His statue stands next to the Museum.

pupil and successor, and between 1919 and 1934 served as professor and director of the Department.

József Béres (1920–2006) rose to prominence through his "Béres drops," a formula containing trace elements, recommended to strengthen the human immune system against tumors. The drops became popular, but the authorities initially hindered his activities. The year 1976 was a turning point in his story when a prominent filmmaker, Ferenc Kósa, recorded a documentary about him and his work.

The historian József Antall (1932–1993) was the director of the Semmelweis Museum. He was the first democratically elected Prime Minister of Hungary in 1990 following the political changes between 1989 and 1990. His statue stands next to the Museum.

———•◆•———

Considerations of public health and epidemiology have been in the forefront of Hungarian medicine. During the "happy peacetime," Austria and Hungary were in the same empire and this facilitated mobility and exchange between Vienna and Pest, then, Budapest (in 1873, Buda, Pest, and Óbuda united into Budapest). During that period, many of the departments and clinics of the medical school moved into modern and spacious accommodation, and well over a century later, many are still using the same buildings. Semmelweis University commemorates many of its former professors with statues, busts, and plaques. We present here a sample of these memorabilia.

The Compromise in 1867 between the Habsburgs and Hungary opened the way to unprecedented progress in the training of medical doctors and in most areas of education. A tremendous expansion program took place between 1873 and 1911, primarily along Üllői Avenue.

By the time World War I began in 1914, the expansive development of the medical school had been completed. The war and the ensuing revolutions of 1918 to 1919, followed by the White Terror and the 1920 Trianon Treaty were extremely trying times. In Chapter 3, we have already mentioned the anti-Jewish legislation, known as *numerus clausus*, which severely influenced the situation at the medical school as well. Then World War II arrived with its multiple tragedies. Many of the Jewish medical doctors perished in the slave labor units. Anti-Semitic hatred caused the guards of these units and responsible army officers to prevent incarcerated doctors from helping wounded soldiers in nearby units of the Hungarian Army, who were often left untreated.

The buildings of the medical faculty suffered greatly in 1945 in the battle of Budapest, because Üllői Avenue was a major venue for the fighting. After the war, the ensuing political changes again interfered with university life. Committees

The Rector's office of Semmelweis University at 26 Üllői Avenue, District VIII.

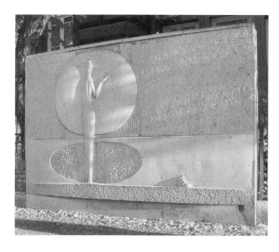

The iron fence encloses a limestone relief by Miklós Borsos, in front of the Rector's office.

examining war crimes eliminated undesired elements, but some innocent people also suffered from careless committee actions. Gradually though the medical school renewed itself under the coming communist dictatorship. Following the suppression of the 1956 revolution and years of ruthless terror, again, a degree of normalcy prevailed and progress was inevitable. Alas, an enormous amount of talent left the country.

The Rector's office building was erected between 1882 and 1884. The interior of the building has recently been renovated and it now shines with its old splendor—although not yet the exterior. The garden with statues, behind the Rector's office building is approachable easily from the Szentkirályi Street side.

Not far from the Semmelweis bust in this garden, there is one in memory of Lajos Arányi (1812–1887). He participated in the War for Independence between 1848 and 1849 and for a while was banned from practicing medicine. He was the first to lecture on pathology at the University of Pest. He was a member of HAS.

Another bust shows Károly Kétly (1839–1927), who was a member of HAS. He specialized in internal medicine.

———◆———

Ágost Schöpf-Merei (1804–1858), a pioneer of pediatric medicine, spent the first part of his career in Hungary; the second in England. Between 1834 and 1849, he

Bust of Lajos Arányi, pathologist.

Bust of Károly Kétly and his relief in the first-floor pantheon of the 2nd Department of Internal Medicine in the inner campus.

taught medical history and pediatrics at Budapest University, and founded a children's hospital. He joined the Hungarian army in the War for Independence and charged János Bokai (see below) with directing his children's hospital in his absence. When Hungary was defeated, Schöpf-Merei went into exile for the rest of his life. He lived and worked in Manchester and founded a children's hospital there, too.

Lajos Markusovszky (1815–1893) was born in Csorba (today, Slovakia). He began his medical studies in 1834 at Pest University. He worked as assistant

EBBEN A HÁZBAN MÜKÖDÖTT
1839-1846-IG AZ ELSÖ MAGYAR
CSECSEMÖ ÉS GYERMEKKÓRHÁZ,
MELYNEK ALAPÍTÓJA ÉS FÖORVOSA
DR. SCHOEPF-MEREI ÁGOST
1 8 0 4 — 1 8 5 8
A MAGYAR TUDOMÁNYOS GYERMEK-
GYÓGYÁSZAT MEGTEREMTÖJE VOLT.

Ágost Schöpf-Merei's bust is in the garden at 74 Üllői Avenue. Memorial plaque on the house at 6 Puskin Street, District VIII. It says that Schöpf-Merei founded the first Hungarian hospital for babies and children and he was its first directing physician. The hospital operated between 1839 and 1846 in this house. He created Hungarian scientific pediatrics.

to János Balassa and spent a few years abroad. In Vienna, he was to become Ignác Semmelweis's friend for life. Markusovszky played important roles in creating the Hungarian military health service during the 1848–1849 War for Independence. He treated the military leader, General Arthur Görgey, and accompanied him into exile after the defeat of the Hungarian cause. When, in 1850, Markusovszky returned home, he was fired from his university job.

Markusovszky was a member of János Balassa's progressive circle. After the Compromise in 1867, Markusovszky joined the educational administration and contributed to the modernization of Hungarian higher education and to the creation of the public health service. He liked to do business informally. His Saturday-evening friendly circle played a significant role in making decisions about new buildings, establishing new departments, and appointing professors. Quality was the number one consideration for Markusovszky, and this is why his circle of friends did not become a clique to hinder progress.

János Bokai (often, Bókay), Sr.'s (1822–1884) bust stands in the garden of the 1st Department of Pediatrics. He was a pediatrician, directed a children's hospital, and initiated new medical procedures. His two brilliant sons in the medical profession, Árpád and János, Jr., changed their surname to Bókay. Árpád (1856–1919) specialized in internal medicine and pharmacology. János, Jr., (1858–1937), like his father, was a pediatrician. Both sons became members of HAS.

In the garden of the clinic at 53 Bókay Street, there is another bust on the right, facing the building, which depicts Sándor Koch (1925–2009), pioneer of

The stone bust of Lajos Markusovszky stands in the garden of the inner campus of Semmelweis University. There used to be another bust, made of bronze, but this has been stolen.

Top left: János Bokai (often, Bókay), Sr.'s bust stands in the garden of the 1st Department of Pediatrics at 74 Üllői Avenue. *Top right*: Árpád Bókay's bust (by Előd Kocsis, 1997) at the entrance of the Bókay Árpád Garden in District XVIII. *Bottom left*: János, Jr.'s bust stands in the garden of the clinic at 53 Bókay János Street, just a block away from his father's memorial. Both of these busts are the work of Béla Radnai (1902).

The bust of virology pioneer Sándor Koch at 53 Bókay János Street, District VIII.

virology in Hungary. In the 1950s, he and his colleagues produced the Sabin drops that were then administered to children in Hungary, saving them from polio.

Toward the city center from Bókay János Street is Leonardo da Vinci Street where there is a relief of Leonardo da Vinci (1452–1519). It contains two quotations. One is by Friedrich Engels, "Leonardo da Vinci was not only a great painter, but also a mathematician and engineer who enriched various areas of physics with his discoveries." The other is by Leonardo: "I can't have enough of it that I serve humankind."

Relief of Lenonardo da Vinci (by Tibor Vilt, 1954), at 52
Leonardo da Vinci Street, District VIII.

Frigyes Korányi (1827–1913) was still a medical student when he participated
in the 1848–1849 War for Independence, after which he completed his training
in Vienna. Only after the Compromise did he succeed in getting a university
appointment appropriate to his qualifications. Eventually, he became one of the
most revered figures in Hungary. He was elevated to a hereditary peerage and was
accorded lifetime membership of the Upper House of Parliament. The Korányis
were a converted and assimilated Jewish family.

He initiated numerous innovations in the medical school, including elevating
the status of laboratory tests and X-ray examinations. He was instrumental in
establishing the tuberculosis sanatorium in Budakeszi (in the green belt close to
Buda) and other improvements in the health service. He was a member of HAS.
His bust stands in the garden of Semmelweis University, at the headquarters
of the Hungarian Academy of Sciences, as well as in the garden of the Budakeszi
Sanatorium. A memorial plaque at the former home of Frigyes Korányi

Frigyes Korányi's busts in the
inner campus of Semmelweis
University and at the Hungarian
Academy of Sciences.

Tombstone of the Korányi family in the Fiume Avenue National Necropolis (plot Á. J. 45). The list of names includes his two sons. Frigyes Korányi, Jr., a financial wizard, occupied high positions in the financial life of Hungary. For Sándor Korányi, see below.

Below: Sándor Korányi's plaque is on the wall of the spacious hall of the Center of Theoretical Medicine. A similar relief is in the pantheon of the 2nd Department of Internal Medicine in the inner campus.

(56 Erzsébet Boulevard, District VII) says: "He was one of the most outstanding Hungarian physicians; the founder of the Hungarian school of internal medicine; the initiator of the struggle against tuberculosis."

Sándor Korányi (1866–1944) was, like his father, a specialist in internal medicine, in particular, diseases of the kidney. He was a pioneer in applying biochemistry and biophysics within medicine. He was a member of HAS. He was seventy years old in 1936 when he was sent into retirement. The reason was not his age but his Jewish roots, regardless of the prominence of his oeuvre and of the medical dynasty of his family. This was not an isolated case either. The internationally renowned ophthalmologist Emil Grósz and his department met the same fate at the same time.

Endre Högyes (1847–1906) studied medicine in Budapest and became professor of medicine, first in Kolozsvár (today, Cluj-Napoca, Romania), then, in Budapest. He was a follower of Louis Pasteur and Robert Koch at a time when their teachings were not yet universally accepted. He organized the Hungarian Pasteur Institute and Hospital. He was a member of HAS.

There is a memorial plaque on the wall of today's 1st Department of Internal Medicine (Korányi Sándor Street, District VIII), which calls Sándor Korányi "the greatest Hungarian specialist of internal medicine."

Memorial plaque for Sándor Korányi at 42 Váci Street, District V, where he lived.

The tombstone of the Högyes family in the Fiume Avenue National Necropolis (by János Kopits, plot Á. J. 44).

Gyula Dollinger (1849–1937) was another surgeon; he founded Hungarian orthopedics, and initiated the association of Hungarian surgeons and a national cancer committee.

Lajos Tóth (1856–1926) was a physician by training, involved with directing higher education at the ministerial level. At around the turn of the nineteenth century, he contributed a great deal to the development of the faculty of medicine of Budapest University.

Left: Bust of Gyula Dollinger (by Anna Kárpáti, 1960) at 78 Üllői Avenue, District VIII.
Right: Relief of Gyula Dollinger (by Erika Szölőssy) at 13 Kék Golyó Street, District XII.

Bust of Lajos Tóth at 78 Üllői Avenue, in the garden between the 1st Clinic of Surgery and the Clinic of Diagnostic Radiology and Oncotherapy.

Memorial plaque for Géza Krepuska at 4 Reviczky Street, District VIII.

Géza Krepuska (1861–1949) was an ear specialist who was a pioneer of otology in Hungary. In addition to treating his broad circle of patients, he taught the younger generations of specialists in his field.

József Fodor (1843–1901) founded the public health system in Hungary. In addition, he did pioneering work in what is today known as immunology. He was born into a Unitarian family in Lakócsa, a village in southwestern Hungary. For high school, he had to move to Pécs where he attended the Gimnázium of the Cistercian Order. He began his medical studies in Vienna and completed

József Fodor's memorial on Gutenberg Square, District VIII.

them in Budapest. He graduated in 1865 and soon he qualified for specializations in surgery, ophthalmology, and obstetrics. He was appointed assistant professor at the Department of State Medicine of Pest University (soon to be the University of Budapest), which included the area known today as forensic medicine.

From the beginning of his career, Fodor was interested in public health and infectology. He took on the duties of chief pathologist at St. Roch Hospital in addition to his university position. He became one of Lajos Markusovszky's protégés. Markusovszky was already engaged in reorganizing the Hungarian health system and medical training. He arranged a prolonged study trip abroad for Fodor. Upon Fodor's return, he was appointed professor of the newly established Department of Public Health Science at Budapest University. He created the scientific foundations of public health and the continuing education of medical doctors; he also initiated the system of school doctors. He was broadly recognized for his achievements, he served as Rector of Budapest University, and he was a member of HAS.

Fodor was much concerned about longevity and noted the conspicuously shorter lifetimes of the Hungarian population in comparison with Western Europe, and even as compared with the twin nation Austria. Hungarian scientists on average lived 51.5 years, their English counterparts lived 67.9 years, and the German scientists 68.3 years. In 1885, he gave a talk at the Hungarian Academy of Sciences on longevity and declared[3]:

> We can live to old age if we live according to the principles of hygiene. We must want to avoid aging. We must have the willpower of not feeling and overcoming the weakening of our body. Genuine belief that we can live longer than what we are witnessing around us, will keep us strong. We need also imagination and a spirited soul. Despair makes one ill; it facilitates aging and paves the way to the grave. Likewise, belief in and trust of the endurance of life and determination to conduct an active life—in addition to a hygienic way of living and peace of mind—are the surest ingredients for a long life.

Pál Heim (1875–1929) was born in Budapest. In 1897 he graduated from the Budapest University Medical School. He specialized in pediatrics and played a pivotal role in creating a national health service for protecting children's health, including the training of nurses. He did research on a variety of children's health problems and carried out extensive publishing activities. Today, a children's hospital preserves his name.

Two venues outside the two principal campuses of Semmelweis University will now be introduced, and a more recent addition. The building at 38 Tűzoltó Street, District IX, houses two departments, namely, the Department of Anatomy, Histology, and Embryology and the Department of Human Morphology and Developmental Biology. This building has traditionally housed two departments with changing names. With growing specialization, departmental designations have become more involved.

Back in the nineteenth century, Géza Mihalkovics (1844–1899), like many of his peers, was as one person a specialist in surgery, ophthalmology, and obstetrics. He did research and treated patients in a variety of areas, and was a much appreciated professor and scientist, and a member of HAS. His bust is in front of the building. Some of his successors have memorials within the building, only a few of which will we mention.

A plaque commemorates Mihály Lenhossék (1863–1937) and his grave has a beautiful tombstone at the Fiume Avenue National Necropolis. He was

Pál Heim's tombstone at the Fiume Avenue National Necropolis by György Vastagh, Jr. (plot 36/2-1-26).

Bust of Géza Mihalkovics (by Miklós Ligeti, 1901) at 38 Tűzoltó Street, District IX.

a descendant of a medical dynasty and was Albert Szent-Györgyi's uncle. Lenhossék attended the Gimnázium of the Piarist Order and in 1886 he graduated as a medical doctor from the University of Budapest. The young Mihály received an introduction to research in his anatomy professor father József's department. In 1899, Mihály became professor of anatomy. When Szent-Györgyi lost his father, Mihály took on the role as his guardian. Mihály Lenhossék was a member of HAS. He died in early 1937; later in the same year, his nephew received the Nobel Prize.

Another plaque in the same building commemorates Tivadar Huzella (1886–1950) who in 1932 moved from Debrecen to Budapest. He stressed the importance of biology in medical training. He was an internationally renowned researcher on intercellular matter and introduced methodological innovations. In 1923, still in Debrecen, he published a book about medical sociology whose pacifist and anti-militarist message so angered the authorities that they banned it, and confiscated and destroyed its copies. The book also appeared in German and in French, and the great French humanist Romain Rolland lauded its publication.

Memorial plaque for Mihály Lenhossék at 38 Tűzoltó Street, District IX. The quote says: "The thorough mastery of human anatomy and physiology does not lead us to the slope of materialism, neither does it exclude the existence of a higher inconceivable power: quite the contrary: it makes us admit the limitations of our knowledge and abilities." The tombstone on his grave in the Fiume Avenue National Necropolis (plot 34-4-14) is by Elemér Fülöp.

Huzella was bravely anti-Nazi; as Hungary was moving toward the catastrophe of World War II, he stood up for his Jewish students.

Memorial plaque for Tivadar Huzella and his lecture theater (as of 2012) at 38 Tűzoltó Street, District IX.

On March 19, 1944, Germany occupied Hungary, and Huzella started his next lecture with the following statement[4]:

> I was supposed to continue this course with a discussion of histology of hormonal organs, but current events disturbed my schedule. We can only know for sure that I can still give this lecture, so I've decided to enlighten you with the scientific critique of race theory . . . We cannot be blinded not to see that race biology is a delusion without data, sound arguments, and logical conclusions. We should never forget that all men belong to the same race. Our common ancestry differentiates us from animals and our ability that we live rationally rather than fighting with teeth and nail. We all belong to the category of the wise man, the *Homo sapiens*. Never you deny of being human!

After the war, Huzella stressed his anticommunist stand and was removed from his professorship. He died neglected and in poverty.

János Szentágothai (1912–1994) was a more recent anatomy professor. His memorial plaque carries his statement: "Anatomy, for me, is not merely the unalienable foundation of the totality of medical thought and action. It is also one of the highest order manifestations of the eternal beauty and harmony of matter and existence." Szentágothai was born into a bilingual German–Hungarian family. He attended a German-language gimnázium at Damjanich Street in Budapest. He studied in Budapest and graduated in 1936. He had already begun to do research during his student years under Mihály Lenhossék.

Regardless of the changing political systems, Szentágothai's career was straightforward—a rare phenomenon. He was an internationally recognized expert on anatomy and was one of the pioneers of so-called functional anatomy, which considered the parts of the human organism as components in action rather than as static details of machinery. His interests encompassed a broad variety of medical topics, and during the latter part of his career, he worked on understanding the functioning of the human brain.

Left: János Szentágothai's image and his ars poetica on a memorial plaque at 38 Tűzoltó Street, District IX. *Right*: János Szentágothai's image as a relief in the Center of Theoretical Medicine, Szentágothai János Square, District IX.

Szentágothai was a Member of Parliament in the last parliamentary cycle under socialism (1985–1990), and in the first parliamentary cycle (1990–1994) following the first free elections after the political changes in 1989 to 1990. For a long time, 1976–1985, he served as President of the Hungarian Academy of Sciences. On the centennial of his birth, his bust was unveiled at the Hungarian Academy of Sciences. Another bust was unveiled in the inner campus of Semmelweis University.

Szentágothai had a critical mind and he lamented the events of the early 1990s. As things have developed further in the years since 2010, one wonders how he would have reacted to them. This is what he said in the early 1990s[5]:

Relying on my modest international outlook, I cannot understand the resurrection of the obsolete and discredited false nationalism and the self-pitying manifestation of one's grievances of Hungarian existence with such an elementary force. This is while they are forcing the other side's liberal ideals that are truly so close to me into political mud wrestling. . . . There emerges, again, anti-Semitism, not yet fifty years after the annihilation (except for some survivors) of the (religious-ethnical) minority that most fortunately used to augment—exactly for its being different—the creative talent of our people in the areas of the economy and culture. Furthermore, they would demand from the gypsies after centuries-long exclusion that they display the same demeanor as the Hungarian, Slav, German, and (in spite of disadvantages, even) Romanian peasant of millennial village culture. . . . We are permitting a new adventurer class with ill-acquired wealth to embark on business.

Emil Grósz (1865–1941), an internationally renowned ophthalmologist, directed the 1st Department of Ophthalmology between 1905 and 1936. He gave the pathological description of several illnesses of the eye and worked out their therapy. His procedure for the surgical treatment of cataracts was applied internationally. Between 1903 and 1918, the government charged him with fighting endemic trachoma at the national level, and he was successful in this undertaking. His bronze bust stands in a secluded office at the Department of Ophthalmology in the inner campus.

János Szentágothai's bust (by Mária Törley, 2012) stands in the Semmelweis Garden.

Emil Grósz's bust is in a secluded office of the Department of Ophthalmology.

Albert Schweitzer's bust, by András Sós, in the Semmelweis Garden.

Two international scientists are commemorated in the inner campus, Albert Schweitzer (1875–1965) and Galileo Galilei (1564–1642). Above we have already referred to the pantheon of professors of medicine on the first floor of the 2nd Department of Internal Medicine. The Department houses several other art pieces beautifully displayed on the walls of its interior.

Another international scientist, the Spaniard Santiago Ramón y Cajal (1852–1934), was honored by a bust in 2013 on Szentágothai János Square, in front of the Center of Theoretical Medicine. Ramón y Cajal's research area was in neuroanatomy and neurophysiology, and he received a share of the 1906 Nobel Prize in Physiology or Medicine for his discoveries in the

Galileo Galilei's relief by Katalin Staindl at the 2nd Department of Internal Medicine.

structure of the nervous system. Opposite Ramón y Cajal's bust, János Szentágothai's new bust was unveiled in a joint ceremony in remembrance of the two scientists.

Busts of Santiago Ramón y Cajal (*left*) and János Szentágothai (*right*) (both by Éva Freund, 2013) on Szentágothai János Square, District IX.

There is a popular statue expressing a vague biological concept of the "Sentry Cell" also in front of the modern Center of Theoretical Medicine of Semmelweis University. It stands on the ground without a pedestal, and conveys a feeling of intimacy. Its curves fortunately augment the straight lines of the modern building behind it.

Two views of the "Sentry Cell" (by Zsófia Farkas, 2010), Szentágothai János Square, District IX.

Center of Theoretical Medicine on Szentágothai János Square, District IX.

Entering the building of the Center of Theoretical Medicine, there is a spacious hall with reliefs of the greats in Hungarian medicine. We have seen above three of them, Albert Szent-Györgyi, Sándor Korányi, and János Szentágothai. Here, more follow in chronological order.

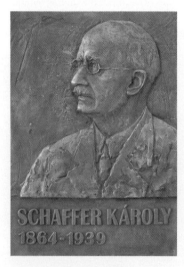

Károly Schaffer (1864–1939), neurologist and psychiatrist, investigated brain tissues and the heritable diseases of the nervous system and mental disorders.

Lajos Nékám (1868–1957), professor of dermatology, initiated the screening of veneral and skin diseases.

Ödön Krompecher (1870–1926), pathologist, did cancer research.

Tibor Verebély (1875–1941), professor of surgery, investigated blood vessels, nerves, the spleen, and stomach ulcers.

Béla Issekutz (1886–1979) founded modern pharmacology in Hungary.

Sándor Mozsonyi (1889–1976), pharmacist and physician, was the first Dean of the Faculty of Pharmacology.

Aladár Soós (1890–1967), surgeon and dietician, introduced the elective hospital diet.

Nándor Ratkóczy (1891–1977), professor of radiology, specialized in the X-ray analysis of stomach and duodenum functions, radiation protection, and radiation therapy.

Left: Imre Haynal (1892–1979) was a specialist for internal diseases and introduced electrocardiography in Hungary. In 1944, as Dean of Medicine of the University of Kolozsvár, he would not allow the Faculty of Medicine to move to Germany. In 1957, as professor of the Budapest medical school, he criticized the communist regime, and had to retire in 1958. *Right*: Gyula Nyírő (1895–1966), psychiatrist and neurologist, researched the pathology of schizophrenia.

József Baló (1895–1979), pathologist, together with his wife, Ilona Banga, a scientist of international renown, discovered a new enzyme in the pancreas, and named it pancreatic elastase. Baló pioneered research on the role of viruses in the development of malignant tumors.

Tibor Nónay (1899–1985), professor of ophthalmology, wrote a monograph about ophthalmological surgery.

Antal Babics (1902–1992), nephrologist, researched diseases of the kidney. He served as minister of health in Imre Nagy's short-lived second government in 1956.

Béla Horányi (1904–1986) was professor of neurology.

József Sós (1906–1973) specialized in the pathology of nutrition and in the so-called diseases of civilization.

Péter Bálint (1911–1998), nephrologist, researched the physiology of the kidney and worked in experimental kidney pathology.

Imre Tarján (1912–2000) was a crystallographer and professor of biophysics.

János Kiszel (1928–1991) was a professor of obstetrics and gynecology, and did pioneering work on the treatment of new-born and premature babies.

László Szabó (1931–2011) was a professor of organic chemistry.

Sándor Juhász-Nagy (1933–2007), professor of medicine, investigated the coronary arteries.

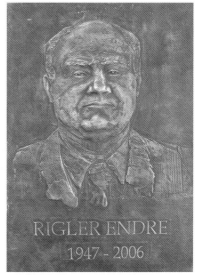

László Romics (1936–2011) was a specialist in internal medicine and researched the metabolism of fatty acids and carbohydrates.

Endre Rigler (1947–2006) specialized in physical education and sport sciences.

NÉMETÚJVÁRI
DR. BATTHYÁNY-STRATTMANN
LÁSZLÓ
1870 - 1931
SZEMÉSZ
A SZEGÉNYEK ORVOSA
EMLÉKÉRE

BATTHYÁNY TÁRSASÁG
BUDAPEST FŐVÁROS TANÁCSA
1989.

Left: Relief of László Batthyány-Strattmann by Éva Oláh Arrè at the Center of Theoretical Medicine. This is a copy whose original was unveiled in Vatican City on the occasion of Batthyány-Strattmann's beatification by John Paul II, in 2003. *Right*: His plaque at 3 Dísz Square, District I.

There is a large relief in this aula depicting László Batthyány-Strattmann (1870–1931) during an ophthalmological surgery. He was an aristocrat who acquired doctorates in philosophy and in medicine, and specialized in ophthalmology. He set up a private hospital on his estate and treated the poor with no charge.

Two figures of Greek mythology decorate the new dental building in Szentkirályi Street. One is Asclepius, the god of medicine, healing, rejuvenation, and physicians in ancient Greek religion. The other is Hygeia, Asclepius's daughter, the

Statues of Hygeia and Asclepius (Péter Párkányi Raab, 2007) decorate the dental school at 47 Szentkirályi Street, District VIII.

Plaque of János Fabini Teofil (1791–1847), the "father of Hungarian ophthalmology," at 9 Zrinyi Street, District V.

Plaque of Imre Magyar (1910–1984), 21b Radnóti Miklós Street, District XIII. He directed the 1st Department of Internal Medicine and was a popular author.

goddess of health and the prevention of sickness. Her name has developed into the English word hygiene.

———— ◆ ————

Beyond the campuses of Semmelweis University, there are additional memorabilia remembering outstanding contributors to the medical profession. A few examples follow.

András Pető (1893–1967) was born in Szombathely, near the Hungarian border with Austria. His father suffered from Parkinson's disease and was confined to a wheelchair. This greatly affected the young Pető's outlook on life along with his mother's adherence to a structured regime for their family's activities. Following World War I, Pető completed his medical studies in Vienna and found employment there in various medical institutions. About his life, legends, rather than facts have spread. Thus, for example, he never accepted money for his medical services; rather, he ran various enterprises to maintain himself financially. He founded and published a magazine and the advertising revenues were his income.

In 1938, after the annexation of Austria by Nazi Germany, the *Anschluss*, Pető returned to Hungary. During the war, Pető hid in the basement of his friends' house, and played with his friends' daughter, who suffered from cerebral palsy. This experience may have prompted him to work on a promising approach to dealing with children whose lives and the lives of their families were made miserable by this illness. Gradually he succeeded in designing an integrated system of treatment, which he called conductive education.

He dealt with children and, increasingly, with pedagogues who would themselves participate in conveying his methods. He developed an institution in which dedicated personnel could earn a degree which qualified them to act as "conductor." He gave hope to countless children with motor disabilities and to

Left: Memorial plaque (by Antal Czinder, 1997) on the house where András Pető lived, on Balassi Bálint Street, at the corner of Stollár Béla Street and Balassi Bálint Street, District V. He is labeled here as medical doctor, pedagogue, writer, and the founder of conductive education and conductor training. *Right*: András Pető's bust (by Viktor Kalló, Sr., 1984) in the hall of the András Pető Institute, 6 Kútvölgyi Avenue, District XII.

their parents that rehabilitation was possible. He overcame all hindrances to win state support for his efforts and his schemes have become a component in the state educational system. The stage was set for his teachings to gain acceptance internationally. Pető died in 1967, but his cause continues.

———•◆•———

We conclude this excursion into the realm of medical memorabilia with an unusual entry. There is a fascinating statue of the writer Frigyes Karinthy at the corner of Karinthy Frigyes Avenue and Irinyi József Street, District XI. It hints at a characteristic approach by Karinthy as though his two selves were conducting a dialog. Frigyes Karinthy (1887–1938) was a most original and sophisticated author, playwright, poet, journalist, and translator. He did some studies in mathematics and physics as well as in the medical arts, but never graduated from college. In 1936, the renowned Swedish brain surgeon, Herbert Olivecrona (1891–1980), operated

Statue of Frigyes Karinthy (by Iván Paulikovics, 1988) at the corner of Karinthy Frigyes Avenue and Irinyi József Street, District XI.

on Karinthy's benign brain tumor in Stockholm.[6] Karinthy's book *A Journey Round My Skull* narrates his experience in detail, and has been appreciated even in medical circles. Karinthy dedicated his book to the noble and true science that has never been so impatient toward superstition as superstition has been toward science. The American neurologist and author Oliver Sacks noted that the book, "the first auto-biographical description of a journey inside the brain, remains one of the best."[7]

———— ◆ ————

We present here two architectural gems that are not part of medicinal sciences, but are in the viciniy of medical departments. One is the headquarters of the Sza-bó Ervin Library and the other is the Museum of Applied Arts. The library began its operations in 1904. Ervin Szabó was a social scientist and one of the initial

Left: The headquarters of the Szabó Ervin Library of Budapest, 1 Szabó Ervin Square, District VIII. *Below:* The Museum of Applied Arts at 33–37 Üllői Avenue, District IX.

Photograph courtesy of József Varga, Budapest.

employees of the library. In 1946, the library was named after him. The headquarters of the library (the center of a whole network of libraries) acquired its present building in 1931; it was the former mansion of the aristocratic Wenckheim family. The other is The Museum of Applied Art at 33–37 Üllői Avenue, District IX, whose founding dates back to 1872. The present building was designed by the architect Ödön Lechner and it was opened in 1896.

Ödön Lechner (1845–1914) was an internationally renowned representative of the Hungarian Secessionist style, which was the Hungarian version of Art Nouveau and Jugendstil. The statue (by Ferenc Kende, 1936) stands in the front garden of the Museum of Applied Arts. The bust (by Sándor Nagy, 1977) stands on Margaret Island.

Notes

1 There is an extended Semmelweis literature. We mention here Endre Czeizel's volume, *Tudósok, gének, tanulságok: A magyar természettudós géniuszok családfaelemzése* (Scientists, Genes, Lessons: Genealogical analysis of Hungarian Scientist Geniuses, Budapest: Galenus, 2006), "Semmelweis Ignác," pp. 18–69.

2 Ignaz Semmelweis, *The Etiology, Concept, and Prophylaxis of Childbed Fever* (translation [from the German original] by K. Codell Carter, Madison, WI, and London, England: University of Wisconsin Press, 1983), p. 129.

3 Czeizel, *Tudósok, gének, tanulságok*, p. 84.

4 I. Hargittai, "Tivadar Huzella: The Man and the Scientist." In Endre A. Balazs and Magdolna Hargittai, Eds., *Tivadar Huzella: Scientist and Humanitarian* (Edgewater, NJ: Matrix Biology Institute, 2012), pp. 22–92.

5 Czeizel, *Tudósok, gének, tanulságok*, p. 350.

6 The operation was successful; Karinthy died two years later of a stroke.

7 Frigyes Karinthy, *A Journey Round My Skull* (New York: New York Book Review of Books, 2008; translated by Vernon Duckworth Barker). Oliver Sacks wrote the Introduction.

5
Pál Kitaibel
AGRICULTURAL SCIENCE

Left: Pál Kitaibel, botanist, chemist, and geologist. His relief (by János Andrássy Kurta) is on the wall of the headquarters of the National Meteorological Service, 1 Kitaibel Pál Street, District II. The inscription is his epitaph, "Gaude Hungaria quae talem tulisti" ("Rejoice, Hungary, that you had such a son"). *Right:* Pál Kitaibel's bust in the Botanical Garden, 25 Illés Street, District VIII.

Pál Kitaibel (1757–1817), a polymath, worked at the medical faculty of the University of Pest and researched in chemistry, mineralogy, hydrology, and especially botany; he was also director of the Botanical Garden. We have chosen Kitaibel primarily for his botany, to symbolize this chapter concerned with the memorabilia of agricultural science.

Kitaibel studied the flora, the minerals, and the hydrography of Hungary and co-authored a monograph about rare plants. In 1789 he discovered a new element, which he did not name, and which later became known as tellurium. When Kitaibel learned that Franz-Joseph Müller von Reichenstein, an Austrian mineralogist, had already discovered this element in 1782, Kitaibel recognized that the credit for the discovery belonged to Müller. The Austrian scientist found the new element while working in Transylvania, but could not identify it. He called it *aurum paradoxium* and *metallum problematicum*.[1] Müller sent a sample of the substance to the German chemist Martin Heinrich Klaproth who isolated the new element, and named it tellurium after the Latin *tellus*—earth.

In Budapest there are six especially noteworthy locations with memorabilia related to agricultural science. They are the veterinary school, the school of gardening and food industry, the ministry for agriculture, the Museum of Agriculture, the Zoo, and the Botanical Garden. We will be referring to these institutions by their popular names rather than their official names which have changed frequently over the decades.

Veterinary School

Human medicine and veterinary science shared the same roots, but their development and organization diverged early on. Joseph II ordered the introduction of the teaching of veterinary science at medical schools and he established the veterinary school in Pest. The first department for the training of veterinarians was formed in 1787 within the Medical Faculty of Pest University. This was soon after the appearance of the first veterinary schools internationally. A Viennese professor, Johann Amadeus Wolstein, was charged with starting the instruction. Wolstein then singled out Sándor Tolnay (1747–1818), and arranged for his education in Vienna. Tolnay went on an extensive study trip to Paris, London, and elsewhere before, in 1787, his department began its operations in Pest. In 1851, the department separated from the medical school and during the ensuing decades, its name and affiliation have changed repeatedly. From 2000, it has functioned as the Faculty of Veterinary Science of St. István University. This university has its headquarters in Gödöllő, a town about thirty kilometers (twenty miles) to the northeast of the outskirts of Budapest. However, the veterinary school has always operated in Budapest and it has been the only school of veterinary science in Hungary.

The principal campus of the veterinary school developed between 1871 and 1881 at the edge of the medium-sized city of about 300,000 that Budapest was at that time. Today, it is still there, at 2 István Avenue, District VII. The campus is squeezed in in the midst of a busy inner city with all the advantages and difficulties of such a location. It occupies an entire city block, bordered by István Avenue, Rottenbiller Street, Dembinszky Street, and Bethlen Gábor Street. Its square shape has a slight truncation at its eastern corner where István Avenue and Bethlen Gábor Street meet. The chopped-off corner is one fourth of the Bethlen Gábor Square.

The campus is an oasis among its surroundings and it is rich in busts and other memorabilia. Much of the further development of the veterinary school has taken

Left: Memorial plaque honors Sándor Tolnay, pioneer of veterinary education at Pest University. It stands on the campus of the veterinary school. *Right:* Bust of Tolnay Sándor (by István Marosits, 1992) under the arcades of the ministry of agriculture (see below).

Statue of Hungarian grey longhorn at the entrance of the veterinary school.

place elsewhere. This campus could have not accommodated, for example, a large-animal clinic, in the middle of a city with a population close to two million people with all the health hazards of such a clinic (except when a large animal appears as a bronze statue).

Over the decades, Hungarian veterinarians have contributed to the development of their science internationally. Ferenc Hutÿra (1860–1934) and József Marek (1868–1952) authored comprehensive texts for training veterinarians. Hutÿra showed that classical swine fever was a viral disease and made a vaccine to protect against it. Marek discovered a disease in chickens, later called Marek's disease. Both were members of HAS. Aladár Aujeszky (1869–1933) was a professor of bacteriology at the veterinary school. He was the first to describe the cattle disease pseudolyssa (also known as Aujeszky's disease) and to distinguish it from rabies.

Relief portraits of four international scientists grace two red-brick buildings, with two portraits on each. They symbolize not only the international character

The statue of centaur (by Attila Boros, 2013) stands close to the entrance of the campus. It honors the 125 years of Hungarian Service of Animal Healthcare. The centaur is part human and part beast; it symbolizes the contribution of veterinary science to human welfare.

The bust of Ferenc Hutÿra on the campus of the veterinary school (by József Damkó, 1937).

Busts of József Marek at the veterinary school (by Ferenc Medgyessy, 1954) and under the arcades of the ministry of agriculture (by Judit Englert, 1978).

Aladár Aujeszky's bust at the veterinary school (by Imre Veszprémi, 1969).

of science, but also the importance of broad foundations in veterinary science. None of the four was a veterinarian; rather, they contributed significantly to chemistry, physics, and physiology.

We introduce the rest of the busts and plaques in the order of the year of birth of those they commemorate.

From left to right: John Dalton (1766–1844), English scientist, pioneer of modern atomism; Antoine-Laurent de Lavoisier (1743–1794), French chemist, pioneer of scientific chemistry; Claude Bernard (1813–1878), French scientist, pioneer of experimental physiology; Hermann von Helmholtz (1821–1894), German physicist and physiologist; honorary member of HAS.

Entrance to one of the buildings on the campus of the veterinary school displaying John Dalton's and Antoine-Laurent de Lavoisier's relief portraits.

Vilmos Zlamál (1803–1886) was a professor at the veterinary school and a member of HAS. He contributed to the introduction of the first legislation on animal health protection in Hungary.

Béla Tormay (1839–1906), professor of animal husbandry and a member of HAS (by Jenő Grantner, 1966).

Lajos Thanhoffer (1843–1909), physician and a member of HAS (by László Garami, 1968). He taught at the veterinary school and at the medical school.

Béla Nádaskay (1848–1933), anatomist and librarian at the veterinary school (by József Damkó, 1938).

Ákos Azary (1850–1888), physician and state veterinarian (by Ede Telcs, 1905).

István Rátz (1860–1917), animal pathologist and a member of HAS.

Gyula Magyary-Kossa (1865–1944), physician, historian of medicinal science, and pharmacologist (by Béla Domonkos, 1990). He was a member of HAS and taught at the veterinary school and at the medical school.

István Bugarszky (1868–1941), chemistry professor and a member of HAS (by Walter Madarassy, 1968). He taught at the veterinary school and the medical school.

Ágoston Zimmermann (1875–1963), veterinarian and a member of HAS (by Walter Madarassy, 1975). He taught anatomy and internal medicine.

Gyula Gróh (see also Chapter 3) was a distinguished professor of chemistry at the veterinary school and he developed a good research laboratory. When in 1919, George de Hevesy was dismissed from the University of Budapest (see Chapter 3) de Hevesy continued his research in Gróh's laboratory. They published a paper, jointly, on the discovery and applications of the radioactive tracer method. From the veterinary school, Gróh moved to the Budapest Technical University and then on to the University of Budapest.

Gyula Gróh (by Tamás Fekete, 1986).

Károly Jármai (1887–1941), veterinarian and specialist in animal pathology (by Kristóf Kelemen, 1986).

Sándor Kotlán (1887–1967), a professor and librarian at the veterinary school (by Richárd Török, 1987).

Rezső Manninger (1890–1970), professor at the veterinary school and a member of HAS (by Kristóf Kelemen, 1990).

János Mocsy (1895–1976), professor at the veterinary school and a member of HAS (by Előd Kocsis, 1995).

Jenő Kovács (1910–1990), veterinarian and chemist, professor of veterinary pharmacology (by Béla Domonkos, 2003).

School of Gardening and Food Industry

Higher education in gardening has a rich history reaching back to as far as 1853. Ferenc Entz founded a gardening school, which developed over ensuing decades, but its status among the institutions of higher learning remained uncertain. In 1939, the school acquired the status of a college and the name Hungarian Royal Academy of Gardening. From 1946 to 1953, the institution operated as the Faculty of Gardening and Ampelology of the University of Agriculture. After 1953, it was again an independent college, and from 1968, a university. In 1992, the school became once again a faculty, of the newly organized Budapest Corvinus University. The Buda campus extends over a beautiful green area in the region of Villányi Avenue, Ménesi Avenue, and Somlói Avenue, District XI. There are busts in its sculpture garden, which is part of the Buda Arboretum.

Bust of Ferenc Entz (1805–1877), medical doctor, gardener, viniculturist, and a member of HAS. The bust (by József Lajos, 1963) stands in front of the library named after him.

János Lippay (1606–1666) introduced the Hungarian language of professional gardening. The statue is by József Seregi (1982).

Máté Bereczky (1824–1895) worked on improving fruit trees (by Ede Mayer, 1898).

István Győry (1861–1954), pharmacist, taught chemistry and his main interest was in fruit production and food processing.

Mátyás Mohácsy (1881–1970), a gardener and author of literature on pomology; he founded large-scale fruit production (by György Jovánovics, 1976).

Gyula Magyar (1884–1945), botanist and gardener. He worked on the improvement of plants.

Imre Ormos (1903–1979); his interest was in garden design and development (by Róbert Csíkszentmihályi, 2003).

Károly Vas (1919–1981), a chemical engineer and internationally renowned food chemist was a member of HAS (by Róza Pató, 1889).

The main building of Corvinus University, just south of the Pest bridgehead of Szabadság Bridge, at 8 Fővám Square, District IX.

The campus in Buda houses three faculties of Corvinus University, namely, food science, horticultural science, and landscape architecture. The headquarters of Corvinus University is on the Pest side; its venue is the old Customs House of Budapest. Miklós Ybl designed it in 1870 and the building opened in 1874. In 1948, the Hungarian government decided to convert it into a venue for higher education, the University of Economics. Instruction started in the academic year 1950/1951. For some time its name was the Karl Marx University of Economics and there is still a Marx statue in the assembly hall of the main building.

Karl Marx (1818–1883) was a German philosopher and economist. He was one of the chief theoreticians of the workers' movement and he was the eponym of Marxism. On July 12, 1989, US President George H. W. Bush gave a speech to the students in the assembly hall of what was still the Karl Marx University of Economics. Mr. Bush stood in front of the Marx statue, but the arrangement of the press platform was such that it hid the statue.

Statue of Karl Marx in the assembly hall of Corvinus University (by Aladár Farkas, 1957).

There is an Imre Nagy relief and memorial plaque in the same assembly hall. Imre Nagy (1896–1958) was a politician and economist of agriculture. He worked out the agricultural policy of the communist regime in the post-World War II period, but protested many of the excesses of the implementation of this policy. He was prime minister of Hungary after Stalin's death, from 1953 to 1955, and presided over a policy of relaxation. In 1955, he was removed from political leadership. In 1956, he returned, and during the revolution, he was prime minister again. In 1958, he was tried and executed. He was a member of HAS, but when he was defeated by the hardline communist leaders, he was stripped of his membership. When he returned to political leadership his status as academician was restored, and he was still a member of HAS when he was tried, sentenced, and executed. There is an Imre Nagy statue on Vértanúk Square (vértanúk means martyrs).

In the middle of the long wall of the assembly hall of Corvinus University, there is a relief of Pál Teleki (1879–1941). He was a renowned geographer, specializing in the economy, and a leading intellectual figure in Hungary in the period between the two world wars. He was professor of geography at the University of Budapest and at one time, he was Rector of Budapest Technical University. He was a member of HAS. He was also a politician, and had a decisive impact on Hungarian politics between the two world wars.

Left: Relief of Imre Nagy in the assembly hall of Corvinus University. *Right:* Imre Nagy's statue unveiled in 1996 on Vértanúk Square, District V (creation by Tamás Varga, son of Imre Varga).

Another view of the statue of Nagy Imre on Vértanúk Square.

The infamous *numerus clausus* law (Chapter 3) was legislated in 1920, during Teleki's first term as prime minister (1920–1921). Increasingly brutal anti-Jewish laws were legislated during his second term (1939–1941). Personally, Teleki wanted yet tougher regulations than the laws prescribed.

He was a leading figure in the revisionist movement, which brought temporary successes when Hitler helped Hungary to regain some of the territories lost in the 1920 Trianon Peace Treaty. Teleki arranged for Hungary to join the three-power agreement, tying the country irrevocably to Nazi Germany, Fascist Italy, and Japan on the road to the catastrophe of World War II. At about the same time he concluded an agreement with Yugoslavia pledging eternal friendship with Hungary's southern neighbor. When Hitler was to attack Yugoslavia and demanded Hungary's compliance by letting the aggressor through, Teleki realized the tragic consequences of his policies, and committed suicide.

Relief of Pál Teleki in the assembly hall of Corvinus University.

TELEKI PÁL
1879 – 1941

Agricultural Ministry

As of 2014, this ministry is now called the Ministry of Rural Development, but its name has changed often, and we will refer to it as the agricultural ministry. It stands on Kossuth Lajos Square in downtown Budapest, opposite the Parliament building. The arcades of the building provide an excellent venue for displaying statues, busts, and plaques.

Statue of "Agronomist Girl" (by Árpád Somogyi, 1959) in front of the agricultural ministry.

We now present the memorabilia within the arcades, with the exception of the relief of Ferenc Erdei (Chapter 2), the busts of Sándor Tolnay and József Marek (seen earlier in this chapter), and those of Sámuel Tessedik, János Nagyváthy, Ferenc Pethe, and Sándor Károlyi, that will appear in the next section, together with other busts at the Museum of Agriculture.

The agricultural ministry on Kossuth Lajos Square, District V, opposite the Parliament building.

Left: The arcades of the southern wing of the agricultural ministry. *Below*: The arcades of the northern wing of the agricultural ministry.

Left: Károly Wagner (1830–1879) created the Hungarian terminology for forestry and contributed to the establishment of an economic policy and legislation for forestry (by Gyöngyi Lantos, 1979). *Right*: János Mathiász (1838–1921) was a pioneer in plant breeding, especially for grapes (by Gyula Nyírő, 1978).

Left: Tamás Kosutány (1848–1915) worked in agricultural chemistry, investigating tobacco and wheat, and researching plant metabolism, he was a member of HAS (by János Horváth, 1984). *Right*: Jenő Kvassay (1850–1919), a water engineer, initiated the institution of civil engineering in Hungary (by András Lapis, 2002).

Left: Sándor Cserháti (1852–1909) was an agricultural botanist (by Jolán Humenyánszky, 1983). *Right*: László Baross (1865–1938), Elemér Székács (1870–1938), and Rudolf Fleischmann (1879–1950) were pioneers in agricultural botany (by Jenő Grantner, 1980).

Left: Imre Ujhelyi (1866–1923) distinguished himself in studies on animal husbandry (by Géza Nagy, 1977). *Right*: Elek Sigmond (1873–1939) did soil research and was a member of HAS (by Péter Kaubek, 1983).

Left: Antal Fasching (1879–1931) worked in geodesy (by Nándor Kóthay, 1982). *Right*: Jenő Hankóczy (1879–1939) was a pioneer in agricultural botany (by Magda Hadik Bódy, 1984).

Left: Gusztáv Szabó (1879–1963) contributed to the development of agricultural technology (by János Horváth, 1983). *Right*: István Okályi (1900–1968) specialized in horticultural policy.

Left: Artúr Horn (1911–2003), a member of HAS, researched in animal husbandry, including animal genetics (by Béla Tóth, 2010). *Right*: Ferenc Donáth (1913–1986) specialized in agricultural economics. He suffered severe punishment for his participation in the 1956 revolution.

Walking around the arcades of the agricultural ministry, one is reminded of the events of the Revolution in October 1956. There are iron spheres in the wall, symbolizing killer bullets, and a memorial plaque. There was a peaceful demonstration in the morning of October 25 on Kossuth Lajos Square during which civilians engaged Soviet armored troops in animated discussions. Suddenly, shooting erupted from rooftops, bullets killing and wounding hundreds of innocent victims. To this day, there is no comprehensive account of what happened. The iron spheres filling the bullet holes indicate a small fraction of the shooting, and they are not necessarily indicative of the actual locations of the bullets, but it is a staggering memento of this tragic event.

 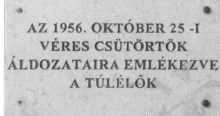

Iron spheres in the wall symbolize bullets killing innocent demonstrators on October 25, 1956 (by sculptor József Kampfl and architect Ferenc Callmeyer, 2001). The plaque says that the survivors remember the victims of "Bloody Thursday" of October 25, 1956.

Museum of Agriculture

Vajdahunyad Castle (in Városliget, District XIV). Ignác Alpár was its architect.

Photo by and courtesy of Michele Hargittai, Duncansville, PA.

The venue of the Museum of Agriculture is exceptional.[2] Toward the end of the nineteenth century, there was unprecedented growth in Hungary in general, and in Budapest in particular. The country was preparing for its millennial jubilee, in 1896, and it was decided to build an ensemble of architectural gems in one place that would reflect the millennial history of Hungarian architecture. From this, the Vajdahunyad Castle emerged, a beautiful and most romantic memento of some of the most famous structures of Greater Hungary. For the 1896 festivities, it was a makeshift construction mainly using wood as the building material. The ensemble became immensely popular and the government decided to build a real structure, under Ignác Alpár's direction. The work lasted from 1904 until 1908, and the result is what we see today. This ensemble houses the agricultural museum.

The main building of the Museum of Agriculture.

Statue and relief honoring Ignác Alpár, the architect of
Vajdahunyad Castle. The statue stands in front of the Castle (by
Ede Telcs, 1931). The relief decorates the Castle.

Ignác Alpár (1855–1928) was an architect, a pupil of Alajos Hauszmann. He
studied in Budapest and Berlin. He aspired to become professor of architecture
at a newly established chair at Budapest Technical University. When he did not
receive the position, he left academia and devoted the rest of his career to design,
in which he succeeded spectacularly. He followed the so-called historicism style,
and determined much of the outlook of today's Budapest with his designs of
many public buildings that date back to the time-period 1867–1914.

We have already seen one of Alpár's designs, the Hungarian National Bank,
where one of its reliefs symbolizes *Science* (Chapter 1). Another relief on the western side of the building depicts the architect himself.

Ignác Alpár's self-portrait is on the western façade
of the Hungarian National Bank (1905). The western
side of the Bank faces Sas Street/Szabadság Square,
District V.

Sándor Károlyi (1831–1906). *Left*: His statue at the Museum of Agriculture, by Alajos Stróbl (1908). *Right*: His bust stands under the arcades of the ministry of agriculture.

In a cozy dent on the northern side of the Vajdahunyad ensemble, there is a statue of Sándor Károlyi (1831–1906), which we show here together with his bust at the ministry of agriculture. Károlyi was a member of the aristocracy, a politician and economist, and a member of HAS. He participated in the War for Independence, 1848–1849, and after its suppression he went into exile in Paris. He returned to Hungary in 1854 and kept himself busy managing his considerable estate. He participated in the regularization of the Tisza River. He developed and advocated a policy of agricultural economics, and promoted the idea of cooperatives.

Relief of King Matthias (see Chapter 1) on the wall near the Károlyi statue.

The statue of the unknown writer Anonymus (Anonymous) stands opposite the Museum of Agriculture. He probably worked for one of the Hungarian kings by the name of Béla and lived some time in the twelfth or thirteenth centuries. He was the author of the oldest written chronicle (in Latin) about the history of Hungary, *Gesta Hungarorum*.

In the early 1930s, a legendary group of young mathematicians formed; they had regular weekly meetings at the Anonymus statue, and the group became identified with this statue. Some of the members of the group emigrated during the late 1930s and

Statue of Anonymus (in Latin; Anonymous in English spelling). The statue (by Miklós Ligeti, 1903) stands opposite the Museum of Agriculture.

had distinguished careers in Australia, the United States, and elsewhere. Others survived the war and continued their teaching and research activities in Hungary. A number of these gifted mathematicians were murdered in concentration camps and in the slave labor camps during the Hungarian Holocaust.

The busts of four scientists stand in the garden of the southeastern side of the Museum of Agriculture. The one honoring Lajos Mitterpacher is the only one that has no counterpart under the arcades of the agricultural ministry.

Detail of the Anonymus statue shows his writing—the beginning of *Gesta Hungarorum*—from around 1200.

Sámuel Tessedik (1742–1820); the bust on the left (by Iván Szabó, 1971) is at the Agricultural Museum; on the right (by János Konyorcsik, 1977), at the agricultural ministry. Tessedik taught agriculture and he was engaged in soil improvement, the introduction of new plants in Hungary, and improving the overall knowledge of agriculture among the general population.

János Nagyváthy (1755–1819); the bust on the left (by László Vastagh, 1960) is at the Agricultural Museum; on the right, at the agricultural ministry. Nagyváthy authored the first Hungarian-language scientific work in agriculture. He initiated the "Georgikon" school in Keszthely for providing scientific training in agriculture.

Ferenc Pethe (1762–1832). The bust on the left is at the Agricultural Museum; on the right, at the agricultural ministry (both busts are by László Vastagh, 1971 and 1977). Pethe edited the first periodical in economics in Hungary and recognized the importance of business organization.

Lajos Mitterpacher (1734–1814) taught agricultural science, geography, and economics at the University. He advocated modern techniques for Hungarian agriculture (by Ferenc Takács, 1962).

Bela Lugosi—"Dracula"—in American horror films; this bust is in a corner niche of the building of the Museum of Agriculture.

There is a fifth bust in a niche of the southeastern corner of the museum building. It depicts the famous Hungarian–American actor Bela Lugosi (1882–1956). He is best known for his roles in horror films, including the character Dracula he played in 1931.

The statues of George Washington (by Gyula Bezerédi, 1906) and Winston Churchill (by Imre Varga, 2003) near the Museum of Agriculture; the Washington statue is on Washington Walkway and the Churchill statue on Churchill Walkway, in the City Park, District XIV.

Close to the Museum of Agriculture in City Park, there is a George Washingon monument, more than a hundred years old. A journalist, Tihamér Kohányi, initiated the project and Hungarians living in America collected the funds. The statue survived all of the political upheavals of the twentieth century. Well after the political changes, in 2003 a bust was erected honoring Winston Churchill at a stone's throw from the Washington monument.

———•◆•———

There is a modest memorial stone dedicated to a foreign botanist, the Swede, Carl Linnaeus—Carl (or Karl) von Linné—(1707–1778), in the largest public park in Budapest called Népliget (Public Gardens), District X. Linnaeus created modern taxonomy (hierarchical classification of biological organisms) for plants and was an early ecologist.

———•◆•———

One of the statues in the garden of the Hungarian National Museum (Chapter 2) is devoted to Ottó Herman (1835–1914), ethnographer, ornithologist, archeologist, and anthropologist. He advocated the importance of science education. There is a memorial plaque in Herman Ottó Street which remembers him.

Left: Ottó Herman's statue (by János Horvay, 1930) in the garden of the Hungarian National Museum. *Right:* Memorial plaque for Ottó Herman at 2 Herman Ottó Street, District II. The plaque says that he was a scientist; wrote about Hungarian ethnography, archeology, and zoology; and founded the Institute of Ornithology.

Imre Wellmann (1909–1994) was a renowned scholar of the history of Hungarian agriculture. He was elected to HAS in 1945, his membership was degraded in 1949, and was reinstated in 1989. The plaque is on his former home at 12 Torockó Street, District II.

Budapest Zoo and Botanical Garden

Entrance to the Budapest Zoo and Botanical Garden, at 6–12 Állatkerti Boulevard, District XIV.

In 1865, the Austrian zoologist, Leopold Fitzinger, founded the Budapest Zoo and Botanical Garden. He was its first director, but public opinion did not like to see a foreigner in this position, and Fitzinger resigned even before the official opening. János Xantus was the first real director from the time of opening. The main entrance of the Budapest Zoo was built in 1912, according to a design by Kornél Neuschloss-Knüsli. This institution has provided tremendous value in education and the dissemination of biological knowledge. To the right of the main entrance there is a small sculpture garden with five busts of former directors.

Left: Former directors of the Budapest Zoo: János Xantus (1825–1894, by Pál Pátzay, 1968). *Right*: Károly Serák (1837–1905, by Péter László, 1989).

Left: Adolf Lendl (1862–1942, by Lajos Petri, 1965). *Right*: Herbert Nadler (1883–1951, by Péter László, 1989).

Csaba Anghi (1901–1982, by Barna Búza, 1986).

Botanical Garden

The Botanical Garden of Eötvös Loránd University is at 25 Illés Street, District VIII, adjacent to the outer campus of Semmelweis University.[3] This garden will be familiar to those who have read Ferenc (Franz) Molnár's famous book for young adults, *Paul Street Boys*, first published in 1906 and which has been in print ever since.[4] Some of the most memorable scenes of the novel take place in the Botanical Garden. In the Garden, there is a conspicuous ivy-clad building with two niches for busts and a memorial plaque. The two weather-worn busts are hard to recognize, but probably depict two ancient authors or philosophers. The plaque commemorates the anti-Nazi politician Endre Bajcsy-Zsilinszky, who on November 23, 1944, was dragged from this building where he was hiding with friends. He was murdered by the Nazis.

The ivy-clad building of the Botanical Garden and the plaque honoring Endre Bajcsy-Zsilinszky.

Busts of two ancient authors or philosophers.

The origin of the Botanical Garden relates to the history of the University.[5] In 1769, the Faculty of Medicine was established, which included the natural sciences in addition to medicine. Jakab Winterl, professor of chemistry and botany, initiated the first botanical garden in 1771, and Pál Kitaibel was appointed its director as Winterl's successor. A photograph of Kitaibel's statue in today's garden served as one of the two chapter-opening images. The Botanical Garden moved to its present location in 1847 due to Palatine Joseph's magnanimous support and he is honored by a memorial column. Originally the Garden occupied a considerably larger area than it does today, but it was reduced just before World War I due to the expansion of the Faculty of Medicine.

A memorial column remembers jointly Sámuel Diószegi (1761–1813) and Mihály Fazekas (1766–1828). The two together edited the first botany monograph in the Hungarian language, *Magyar fűvészkönyv* (Hungarian Herbal), published in 1807. Diószegi was a Calvinist minister and botanist. Fazekas, Diószegi's brother-in-law, is best known for his epic poem *Lúdas Matyi* (Mattie the Goose-boy). It is about a popular folk hero who struggles against and then ends up victorious over his unjust lord.

There is a memorial to former director Lajos Jurányi (1837–1897), a member of HAS, who was directior of the garden in the period following the Compromise; the garden reached the summit of its development in his charge. The botanist member of HAS, Rezső Soó (1903–1980), excelled as director after World War II in reconstructing the garden and alleviating its heavy losses.

Memorial columns for (*left*) Palatine Joseph and (*right*) Sámuel Diószegi and Mihály Fazekas.

169

Memorials honoring (*left*) Lajos Jurányi and (*right*) Rezső Soó.

Notes

1 This resembles Albert Szent-Györgyi, when he discovered a new substance that he would eventually identify as Vitamin C, first he called it "godnose," meaning that nobody knew though God might.

2 The website of the museum: <http://www.mezogazdasagimuzeum.hu>.

3 The website of the Botanical Garden: <http://www.fuveszkert.org>.

4 Ferenc (Franz) Molnár (1878–1952) was a Hungarian–American novelist and playwright.

5 <http://www.fuveszkert.org/a-fuveszkert-tortenete/> (as of the end of 2013).

6

Theodore von Kármán

ENGINEERING AND INNOVATION

Theodore von Kármán lecturing at the California Institute of Technology.
Courtesy of Roger Malina, Richardson, TX.

Theodore von Kármán (1881–1963) is one of the great names among the graduates of the Budapest University of Technology and Economics (BUTE). He was the only "Martian" (more about the five Martians in Chapter 7) who completed his university education in Hungary. In 1902, he graduated as a mechanical engineer and got his first taste of research under Donát Bánki's (see below) mentorship.

The exact name of the School von Kármán graduated from was the Royal Palatine Joseph Technical University of Budapest, named after the Habsburg Archduke Joseph (1776–1847). After the king, he filled the second highest position of the Land as Palatine—a sort of governor, and he was popular in Hungary.

Palatine Joseph's namesake, King Joseph II (see Chapter 3), ordered that only graduates of the *Institutum Geometricum* could be employed as public engineers, which helped them to find jobs. Most of Joseph II's reforms did not survive him, but the Institutum continued to exist because it was part of the university.

Busts of Theodore von Kármán in the sculpture garden of the
Budapest University of Technology and Economics (by Ilona Barth,
née Mezőfi Mózer, 1994), and (by Gyula Meszes Tóth) in front of the
Museum of Transportation (see below).

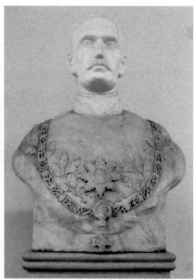

Statue of Palatine Joseph on József Nádor Square, District V, in downtown Budapest
(by Johann Halbig, 1869), and his bust in building "K."

The importance of water economy expressed itself in the name change to
Institutum Geometrico-Hydrotechnicum. Pál Vásárhelyi (1795–1846), its most
famous graduate, became a water engineer and pioneered river control and flood
prevention. He did extensive and fundamental work in regulating the flow of the
Danube River and made similar contributions with far-reaching consequences

Left: Pál Vásárhelyi's bust (by László Szomor) at the Museum of Transportation at Városligeti Boulevard, District XIV. *Right*: His tombstone at the Fiume Avenue National Necropolis (plot 10/2-island; by Jenő Grantner, sculptor, and István Zilahi, architect).

to the Tisza River. He recognized the healing powers and thermal energy content of the underground water reserves of the Great Hungarian Plain. He was a member of HAS.

Ferenc Reitter (1813–1874), another student of the Institutum, participated in the regularization—flood control—of rivers and in building one of the first railway lines in Hungary. From 1851, he was the highest ranking architect in the capital city. He directed the development of the embankments of the Danube and the city planning of Pest-Buda, which was not yet Budapest. These plans contributed to the emergence of the modern metropolis on the two banks of the Danube after the 1873 unification of Buda, Pest, and Óbuda. Reitter was a member of HAS.

Eventually, the Hungarian nobility—the political elite—recognized the obsolete state of the country as compared with the developed Western Europe, and the era of reforms followed, between 1825 and 1848. In 1836, Parliament discussed a proposal to establish a *Politechnicum* for training young people in commerce and manufacturing, as well as in mathematics, physical sciences, geometry, and mechanics. In 1844, King Ferdinand V decreed the establishment of the School

Memorial plaque at 1 Reitter Ferenc Street, District XIII. The text emphasizes Reitter's merits in developing the sewerage system of Budapest and in designing the embankments of the Danube.

of Industry, in whose development István Széchenyi (see Chapter 2) played a pivotal role. The location of the rector's office of Eötvös University of today (see Chapter 3) was the first venue for the Royal Joseph School of Industry (named after the Palatine in 1846).

With increasing recognition of the importance of training engineers, by the middle of the nineteenth century, such training was moved from the University and joined with the School of Industry. The emerging Institute of Engineering found its first home in an impressive building at what is today the corner of József Boulevard and Népszínház Street, District VIII. There is no visible indication on the building that would inform the passer-by as to its historical significance to higher education.

The independent Budapest Technical University came into existence in 1871, and was initially located in rented accommodation at what today is the corner of Gönczy Pál and Lónyai Streets.

The School had the designation "university" in its name from the very beginning. Basic science played an important role in both instruction and in the professors' research activities and was on an equal footing with technology. József Eötvös was the minister of religion and public education at the time and he was

Above: The corner bulding of József Boulevard and Népszínház Street, District VIII, one-time home of the Institute of Engineering. Alajos Hauszmann was its architect. His bust (by Jenő Bory, 1927) stands in the assembly hall of building "K" of BUTE.

Left: The corner of Gönczy Pál and Lónyai Streets, District IX. In 1871, the new Technical University started its operations here.

aware of the significance of the university designation, as shown by the motion he submitted in 1870 to the House of Representatives on this matter[1]:

> Whereas, at the Technical University, mathematical and natural sciences are lectured in at the same high level of research and with the same liberal method of science as all sciences in general at the University; whereas students at the Technical University are required to obtain the same degree of skills and qualifications for an engineering and technical career as, for instance, those preparing for a legal or medical career . . . I find it expedient to announce, by way of a legal act, that the Technical University shall be of the same standing as other state universities and that afterwards it should be organised as a principal educational institution of university rank.

The new university rapidly outgrew its first venue and the solution was therefore to build a new campus. This was in the area between today's Múzeum Boulevard and Puskin Street (Eszterházy Street, then). We showed this in Chapter 3. In 1909, the new central building ("K") of the Technical University was ready and the school moved across the Danube to its present location.

The campus of what is known today as the Budapest University of Technology and Economics (BUTE) is lined with trees and busts of former professors, a

The middle part of the central building "K" of the Budapest University of Technology and Economics, 1–3 Műegyetem Quay, District XI, Alajos Hauszmann's design, opened in 1909.

The four statues at the main entrance symbolize the original faculties: chemical engineering, architecture, general engineering, and mechanical engineering. These statues (by Károly Senyei, 1909) perished in World War II; replacements stand in their place today.

The following artists created the replacement statues after photos of the originals (in the order of the statues): Botond Polgár and Orsolya Klosz; Írisz Kulcsár and Patrik Kotormán; Krisztián Máthé, Tamás Gilly, and Sándor Zoltán Nagy; and András Engler, András Kontúr, and Attila Sólyom. The stand of the statues was designed by Csongor Gábor Szigeti.

The central walk-way of the campus of the Budapest University of Technology and Economics. The bridge at the back connects building "K" (*on the right*) with the library (*on the left*).

number of which will be shown below, along with those in the central assembly hall, the Aula, in building "K," in a loosely chronological order of year of birth.

István Kruspér (1818–1905) was the first dean of the Faculty of Civil Engineering. He was a graduate of Vienna Technical University. He advocated standardization in measurements and was active internationally in establishing the measuring and weighing standards for the kilogram and the meter. He was a member of HAS. József Sztoczek (also, Stoczek, 1819–1890) taught natural science at the Institute of Engineering, and in 1871 he became the first Rector of the Budapest Technical University. His research interest was in the mathematical aspects of physics. He was a member of HAS. Kálmán Szily (1838–1924) was a science organizer in addition to his research interest in the mechanical theory of heat. He served at one time as rector and he was a member of HAS.

Antal Kherndl (1842–1919) studied in Budapest, Germany, and Switzerland; his specialty was bridge construction. Beside teaching and theoretical research, he participated in designing most of the bridges across the Danube. He was a member of HAS. Vince Wartha (1844–1914) was a chemist and long-time rector. He was a member of HAS, and made efforts to enhance the technological sciences at the Hungarian Academy of Sciences. He invented the eozin cover for ceramics. His wife,

From left to right: Three busts in the Aula in building "K," István Kruspér (by Károly Senyei, 1918), József Sztoczek, and Kálmán Szily (by Jenő Bory, 1914).

Vilma Hugonnai, was the first female medical doctor in Hungary (Chapter 4). Sándor Lipthay (1847–1905) was a railway engineer who contributed much to both the theoretical development and the actual construction of railways. He promoted the mathematical preparation of engineers and served as rector. He was a member of HAS.

Gyula Kőnig (1849–1913) studied in Vienna, Heidelberg, and Berlin; he had been interested in medicine and physics before he decided on mathematics. He was a member of HAS and served the Technical University as rector.

Lajos Ilosvay (1851–1936) was a chemistry professor and a member of HAS.

Four more busts in the Aula: *Top left*: Antal Kherndl (by János Pásztor, 1916).
Top right: Vince Wartha (Jenő Bory, 1914). *Bottom left*: Sándor Lipthay. *Bottom right*:
Bust of Gyula Kőnig (by János Pásztor, 1994). Kőnig has a grave in the Fiume Avenue
National Necropolis (plot 10/1-1-12) which is a joint one with his son, Dénes Kőnig (see
Chapter 8).

The friends and admirers of Lajos Ilosvay erected the plaque on the left in building "CH" while he was still alive. His bust (by Jenő Bory, 1937) is in the Aula.

The chemistry building "CH" (Győző Czigler's design, 1904) at 4 St. Gellért Square, District XI.

Antoine Lavoisier.

Georg Lunge.

August Wilhelm von Hofmann.

Michael Faraday.

Jöns Jakob Berzelius.

On the southern façade of the chemistry building "CH," that is, on the back of the building, there are reliefs of five great international scientists. The reliefs show the damage from decades of weather, and two of them are hidden when the trees blossom in front of them. Except for the lesser known Georg Lunge, they are easily recognizable even without reading their names. The first names, incidentally, appear in their Hungarian translation, as was customary at the time.

The most famous chemistry graduate of the Technical University is the Nobel laureate George A. Olah (see also Chapters 1 and 7). He began his research career in a laboratory in building "CH" and had a special setup on one of its balconies because some of the materials he used were explosive.

The Nobel laureate physicist Eugene P. Wigner (see also Chapters 1 and 7) and Edward Teller (1908–2003), known as the father of the hydrogen bomb, were earlier students, both in chemical engineering and both only briefly, Wigner in the academic year 1920/1921 and Teller in 1925/1926. According to the transcripts, both took up additional subjects in addition to those prescribed for freshman chemical engineering students. For Wigner, crystal optics appears, anticipating his thesis research in Berlin and for Teller, the choice of vector analysis and the theory of relativity appear equally forward-looking.

In 1921, Wigner moved to Berlin to continue his studies and upon graduating as a chemical engineer he stayed on to earn his doctorate. He then returned to Budapest and joined the Mautner Tannery where his father was manager.

George A. Olah and one of the authors in 1996 at the University of Southern California.

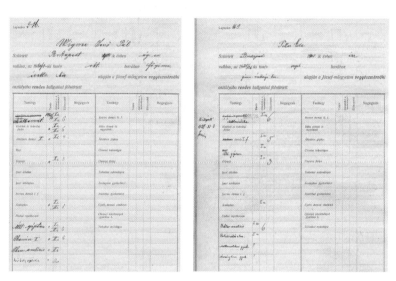

Eugene P. Wigner's and Edward Teller's transcripts at the Budapest Technical University.
The transcripts are courtesy of the Archives of the Budapest University of Technology and Economics.

Eugene P. Wigner's bust on the campus. In January 2005, the 2004 chemistry Nobel laureate Hungarian–Israeli Avram Hershko visited Wigner's bust.

Left: Edward Teller, Béla Pogány, Mrs. Mullikan, and Robert S. Mullikan at the Fishermen's Bastion. Béla Pogány (1887–1943) was a professor of physics whose main interest was in acoustics and optics. He was a member of HAS. *Center*: Robert S. Mullikan, Mrs. Mullikan, Béla Pogány, and Edward Teller in the Castle District of Buda. *Right*: the same location in 2012.

Courtesy of the late George Marx.

Edward Teller visiting John von Neumann's bust on the campus. John von Neumann (1903–1957), pioneer of the modern computer, was one of the Martians (see Chapter 7).

Photo by and courtesy of János Philip, Budapest.

Edward Teller left the Budapest Technical University after the fall semester of the academic year 1925/26, for the Karlsruhe Technical University. When the future Nobel laureate Robert S. Mullikan and his wife visited Budapest in 1928, Teller helped them get around.

Bust of Ferenc Schafarzik (by Tibor Vilt, 1981) on the campus.

Károly Zipernowsky's bust (by Lajos Petri, 1960) in the Aula.

Continuing with the busts on the campus of BUTE, Ferenc Schafarzik (1854–1927) taught geology at the Technical University and conducted extensive research in the area of geological sciences. He was a member of HAS.

Károly Zipernowsky (1853–1942) was one of the first electrical engineers at the Technical University. Ábrahám Ganz put him in charge of the newly established electrical division of the Ganz factory. Later, Zipernowsky became a leading professor of electrical engineering. He was the co-inventor of the transformer and the alternating current generator. Ottó Bláthy and Miksa Déri were the other co-inventors. Zipernowsky's bust is in the Aula. A plaque at 39 Lövőház Street, District II, commemorates all three inventors. They were engineers at the division of electrical technology of the Ganz and Partner Company. Further plaques commemorate the co-inventors. A plaque for Károly Zipernowsky is on the school named after him, at 1–3, Zipernowsky Street, District III; Ottó Bláthy's plaque is at 7 Bláthy Ottó Street, District VIII; and Miksa Déri's is at 1 Déri Miksa Street, District VIII.

Donát Bánki (1859–1922) graduated as a mechanical engineer from the Technical University. For years, he worked at the Ganz Works. He designed motors and improved the carburetor. He was the first to use water injection in internal combustion engines to protect the walls from overheating. In 1899, Bánki became professor of mechanical engineering at the Technical University. Bánki was von Kármán's mentor who helped him embark on his creative research career. There is a memorial plaque where he lived, at 6 Rózsahegy Street, District II.

Bust of Donát Bánki (by Pál Pátzay, 1959) in the Aula.

The engineer Szilárd Zielinski (1860–1924) was the first graduate of the Technical University to earn a doctorate (the plaque refers to this fact). He adopted the use of reinforced concrete in building in Hungary.

Ignác Pfeifer (1867–1941) was a chemical engineer who at one time was in charge of the chemistry institute of the Technical University. Water softening was his area of research. Because of his involvement in the 1919 communist regime, he had to leave the Technical University and he was invited to head the research laboratory of Tungsram, the manufacturer of light bulbs and vacuum

E HÁZBAN ÉLT ÉS ALKOTOTT
1910–1924–IG

ZIELINSKI SZILÁRD
(1860 – 1924)

MÉRNÖK, MŰEGYETEMI TANÁR,
AZ ELSŐ MŰSZAKI DOKTOR,
A VASBETON ÉPÍTÉS HAZAI MEGHONOSÍTÓJA
ÉS A MÉRNÖKI KAMARA ELSŐ ELNÖKE.

MÉRNÖKI KAMARA
ETELE KÖR
BUDAPEST FŐVÁROS–
XI. KERÜLETI ÖNKORMÁNYZAT

1994.

Szilárd Zielinski's bust (by Tibor Zielinski, 1994) on the campus and his plaque at 3 Budafoki Avenue, District XI.

tubes. There, Pfeifer created a modern research organization in Hungary of which he was a pioneer. At the end of the 1930s, the anti-Jewish laws jeopardized his employment at Tungsram.

Aurél Liffa (1872–1956) was a geologist and mineralogist, professor of mineralogy of the Technical University, and director of the Hungarian Institute of Geology. He surveyed ores, minerals, and soils in Hungary, including the occurrence of kaolin. Győző Mihailich (1877–1966) graduated from the Technical University. His field of interest was the structures of reinforced concrete. He was professor and rector, and a member of HAS.

Busts of Ignác Pfeifer (*left*, by Sándor Mikus, 1975) and Győző Mihailich (*right*, by Tibor Vilt, 1983) on the campus.

EBBEN A HÁZBAN LAKOTT
1928.JÚLIUS 1-TŐL 1956.OKTÓBER 23-ÁN
BEKÖVETKEZETT HALÁLÁIG

PROF. DR. LIFFA AURÉL

GEOLÓGUS, MINERALÓGUS, FŐBÁNYATANÁCSOS,
A MAGYAR KIRÁLYI FÖLDTANI INTÉZET
CÍMZETES IGAZGATÓJA, A MÜEGYETEM
ÁSVÁNY-ÉS FÖLDTANI TANSZÉKÉNEK TANÁRA,
A HAZAI ÁSVÁNYOK, ÉRCEK, TALAJOK ÉS
KAOLINELŐFORDULÁSOK KIVÁLÓ KUTATÓJA

SZÜLETÉSÉNEK 130. ÉVFORDULÓJÁN ÁLLITOTTA
A MAGYAR ÁLLAMI FÖLDTANI INTÉZET,
VALAMINT UNOKÁI ÉS DÉDUNOKÁI.
2002.

Plaque of Aurél Liffa at 42 Damjanich Street, District VII.

Farkas Heller (1877–1955) graduated from the law school of the University of Budapest and taught economics at the Budapest Technical University, which he briefly served as its rector between 1945 and 1946. In 1949 his HAS membership was downgraded to an advisory status; it was reinstated in 1991, posthumously.

Géza Zemplén (1883–1956) was a world-renowned organic chemist and a member of HAS. After graduating from the University of Budapest, he spent years with the Nobel laureate Emil Fischer in Berlin. Zemplén's main interest was in carbohydrate chemistry, including the most diverse derivatives of sugars. He had considerable achievements in pharmaceutical chemistry. George A. Olah was among his pupils in Budapest.

Busts of Farkas Heller (by Emil Eőry, 2007) on the campus and of Géza Zemplén in the Aula.

László Verebélÿ was professor, chairman of department, dean, rector, and a member of HAS; he trained electrical engineers. He had broad international experience and participated in the electrification of the railways in Hungary. József Liska (1883–1967) was an electrical engineer who studied at the technical universities of Vienna, Budapest, and Karlsruhe. He worked on the electrification of the Hungarian railway system. From 1914, he taught at the Royal Palatine Joseph Technical University. He was a member of HAS.

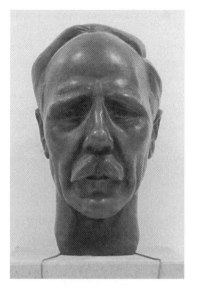

Busts of László Verebélÿ in the entrance hall of building V1 of BUTE and in the Museum of Electro-techniques at 21 Kazinczy Ferenc Street, District VII.

József Liska's bust in the entrance hall of building V1.

Géza Pattantyús-Ábrahám (1885–1956) was a mechanical engineer. He initiated the continuing education of engineers. In addition to his long-time professorial activities, he worked in the industry and contributed much to its modernization, especially in the areas of pumps, transportation, lifts, mining machinery, and enhancing the culture of technology in Hungary. The physicist Zoltán Gyulai (1887–1968) researched the condensed phase. He was an assistant to Béla Pogány and then to Károly Tangl. Gyulai resumed his teaching following a seven-year period as a POW in Russia (1915–1922). He worked as a visiting scientist in Szeged, Göttingen, Debrecen, and Kolozsvár (now, Cluj-Napoca, Romania). Between 1947 and 1962, he was chair of experimental physics at the Technical University. He was a member of HAS.

Busts on the campus, *left*: Géza Pattantyús-Ábrahám (by Péter László, 1983); *right*: Zoltán Gyulai (by Vilmos Szamosi Soós, 2000).

Iván Kotsis (1889–1980) was an architect and a professor, and in 1945, he was elected to the membership of HAS. In 1949, he was stripped of his Academy membership and his professorial appointment was terminated. He engaged in extended architectural design work for the rest of his life. His HAS membership was reinstated in 1989, posthumously. Ádám Muttnyánszky (1889–1976) was a professor of mechanical engineering, much involved in solving practical problems. His students appreciated his pedagogical activities and honored him with a memorial plaque at 8–10 Margit Boulevard, District II.

Busts on the campus: *left*: Iván Kotsis (by Gyöngyi Szathmáry, 1998); *right*: Ádám Muttnyánszky.

József Varga (1891–1956), a chemical engineer and politician, was professor of chemical technology at the Budapest Technical University, later at the University of Veszprém. He was a member of the government between 1939 and 1943, and in 1941, he was the only member of the government who was opposed to Hungary entering the war. His research and industrial activities concerned the utilization of oil and coal.

Bust of József Varga (by István Kákonyi, 1966) in the Aula.

Oil rig model in front of the headquarters of MOL, the Hungarian Oil and Gas Company, at 18 Október huszonharmadika Street, District XI.

Jenő Egerváry (1891–1958) was professor of mathematics and a member of HAS. He continued Dénes Kőnig's (see Chapter 8) pioneering work on the theory of graphs and extended it toward applications in economics. For some time between the two world wars, he was not allowed to teach because he gave lectures in 1919 at the time of the communist dictatorship. János Proszt (1892–1968) was professor of inorganic chemistry and a member of HAS. He investigated the properties of solutions, worked out methods for silicon production, introduced polarography (a new electroanalytical technique) in Hungary, and wrote about the history of chemistry. József Jáky (1893–1950) was professor of transportation and did considerable design work; for example, he participated in developing Ferihegy Airport (today Liszt Ferenc International Airport), the motorway connecting the city with the airport, and the Danube Iron Works.

From left to right: Jenő Egerváry's bust (by Csaba Bodó, 2006) on the campus, János Proszt's in building "CH," and József Jáky's on the campus.

Gusztáv Rados (1862–1942) was a mathematician who studied in Budapest and Leipzig. He was professor of mathematics at the Budapest Technical University and at one time served as its rector. His son, Jenő Rados (1895–1992), was an architect and historian of architecture. He was professor until 1957 when he was dismissed for having served on the Revolutionary Committee of the Faculty of Architecture, in 1956. However, for a long time, he remained active and did design work for companies. The tombstone at the Farkasréti Cemetery lists father and son among the Rados family members.

Pál Csonka (1896–1987) was the son of János Csonka (see below). Pál was professor of materials engineering (as we would call it today). His best known work was his study of shell structures. In 1957, he was dismissed from his professorship because of his participation in the 1956 Revolution; he continued in a research position. Imre Korányi (1896–1989) was a professor and a civil engineer, and did a lot of design work. In 1956, he was a member of the Revolutionary

Top left: Gusztáv Rados's bust (by Jenő Bory, 1954) in the entrance hall of the Library of the Technical University; *Top right*: the tombstone of the Rados family at the Farkasréti Cemetery (plot 7/7-1-9/10); *Bottom left*: Jenő Rados's bust (by Gyöngyi Szathmáry, 2000) on the campus.

Committee of the Faculty of Civil Engineering, and in 1959 he was forced to retire from his professorial appointment. He continued his engineering activities for fifteen more years before properly retiring.

Busts on the campus. *Left*: Pál Csonka (by Gyöngyi Szathmáry, 1996); *right*: Imre Korányi (by Gábor Szabó, 2006).

István Náray-Szabó (1899–1972) was an internationally renowned crystallographer. In the late 1920s, he spent some time with the pioneer of X-ray crystallography, William L. Bragg, in Manchester. Náray-Szabó taught and did research in Szeged and at the Budapest Technical University. In 1945, he became a member of HAS, but was soon expelled when he was arrested, tried, and sentenced to years of incarceration for allegedly conspiring to destroy the Republic. He continued his studies and writing in prison. Afterwards, he built up another strong career in academia.

Géza Schay (1900–1991) was a physical chemist. Between the two world wars, he worked in the State Chemistry Institute. After 1945, he was elected as a member of HAS and for many years he was in charge of physical chemistry at the Budapest Technical University. He also served as director of the new Central Research Institute of Chemistry of the Hungarian Academy of Sciences—one of the predecessor institutions of the current Research Center for Natural Sciences (RCNS) of HAS.

Georg von Békésy (1899–1972) studied at several locations in Europe as the assignments of his diplomat father kept changing. He attended college at the University of Bern and majored in chemistry and physics. For his PhD he did research in physics under Károly Tangl at the University of Budapest. He then found employment at the Experimental Station of the Hungarian Post Office. For a while, he was an odd man out, being a physicist PhD among engineers, but he proved to be very good in problem solving and, as recognition, his institute granted him the title of engineer. He was also recognized as a scientist and, between 1933 and 1946, he taught experimental physics at the University of Budapest; between 1941 and 1946 he was the head of department.

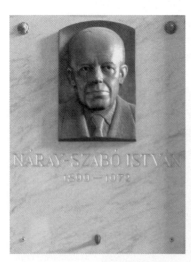

Reliefs of: *left*: István Náray-Szabó and *right*: Géza Schay (1900–1991); both in building "F", where they used to work at the physical chemistry department. There is a similar relief for Schay in the RCNS.

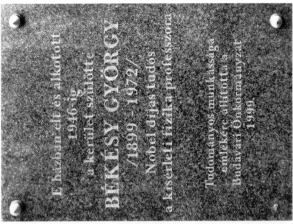

The plaque on this building commemorates the scientific work of the Nobel laureate scientist and professor of experimental physics.

Von Békésy lived in the house in the middle, until 1946, at 19 Fő Street, District I.

The building on the left was Georg von Békésy's birthplace; it gave way to a large apartment house (1 Pauler Street, District I).

The drawing is by Gyula Széles; courtesy of László Kovács, Szombathely.

He never stopped working at the research laboratory of the Post Office. In 1946, he left Hungary, and after a year at the Karolinska Institute in Stockholm, he moved to Harvard University in the United States where he did research until 1966. He spent the rest of his life in Honolulu, Hawaii.

Two views of the abandoned building (as of Fall 2013) where the Experimental Station (later, Institute) of the Hungarian Post Office was located, at Zombori Street, District IX, behind the Puskás Tivadar Technical High School of Telecommunication. There was in the past a plaque there in memory of Békésy.

In 1961, von Békésy received the Nobel Prize in Physiology or Medicine (see Chapters 1 and 3). The presenter of his Nobel award stressed that von Békésy made fundamental discoveries concerning the dynamics of the inner ear, which resulted in clinical applications, and that his discoveries were "entirely the result of one single scientist's work."[2]

It was unusual that a physicist should receive the medical Nobel Prize, and von Békésy gave an unusual Nobel lecture in which half of the illustrations were images of ancient art pieces, which he managed to connect to his science. Von Békésy had amassed a great collection of art pieces, especially of Central-American Indian culture, which he willed to the Nobel Foundation. When we visited the Nobel Foundation at 14 Sturegatan, Stockholm, we saw that magnificent pieces from von Békésy's collection decorated the otherwise spartan offices of the Foundation.

Von Békésy's interest in the mechanism of hearing and the structure of the human ear dated back to the mid-1930s. He began with learning about the intricacies of the anatomy of the human ear. He did not find that the model of the inner ear used to instruct medical students was sufficiently detailed and he turned for help to anatomy professor, Ferenc Kiss. The anatomist presented von Békésy with a human head and von Békésy and a surgeon dismantled half of this head to extract the bones responsible for hearing. At the end of this surgical procedure, almost nothing remained. There was still the other half of the head and von Békésy took this to his own laboratory. There, he dismantled this other half and

Georg von Békésy's bust is at the Puskás Tivadar Technical High School, 22 Gyáli Avenue, District IX.

he arrived at the inner ear, which he examined under the stereo microscope. The beauty of the inner ear amazed him, and he decided to stay with this project.

Von Békésy had already received a great deal of recognition, internationally, long before his Nobel Prize. In 1939, the Hungarian Academy of Sciences elected him a corresponding member. When he left Hungary in 1946, he was stripped of his membership, which was reinstated in 1989, posthumously.

Dennis Gabor (1900–1979) was a Nobel laureate in Physics for the invention of holography (see Chapter 1). He spent a very short time as a freshman at the Faculty of Mechanical Engineering of the Budapest Technical University, and completed his studies in Berlin. After the Nazi takeover, he moved to England where he had a brilliant career. There is more about Gabor in Chapter 7.

Bust of Dennis Gabor on the campus of the University of Technology and Economics.

The bust of Zoltán Bay, at 14 Görgey Artúr Street, District IV—Újpest.

Zoltán Bay (1900–1992) graduated from the University of Budapest in 1923 as a high school teacher of mathematics and physics, and he earned a doctorate in physics in 1926. Between 1926 and 1930, he did research in Berlin. He returned to Hungary and obtained a professorial appointment in theoretical physics at Szeged University. Between 1936 and 1948, he directed the research laboratory of the United Incandescent Lamps and Electrical Company (Tungsram) in Budapest. From 1938, he had a second job as the founding professor of atomic physics at the Budapest Technical University, under Tungsram's sponsorship. Bay helped his persecuted Jewish co-workers between 1944 and 1945. The Gestapo arrested Bay briefly for preventing the Germans from removing Tungsram's equipment. After the war, in 1948, the secret police of the communist dictatorship also arrested him for a short time. In 1949, Bay left Hungary, and built a successful career in the United States. He was research professor at George Washington University and then, group leader of atomic physics at the then National Bureau of Standards (today, it is the National Institute of Standards and Technology). Israel recognized him as Righteous among the Nations and listed his name at the Yad Vashem Memorial. Bay's best-known scientific achievement was that he produced and detected echoes of radar waves from the Moon. Bay was a member of HAS, but the Academy annulled his membership when, in 1949, he left the country. In 1989, his membership was reinstated. Paradoxically, in 1981, he had already been elected as an honorary member of HAS.

Zoltán Csűrös (1901–1979) was an organic chemist and a member of HAS, with strong ties to the polymer and textile industries. In 1938, he founded the

Department of Textile Chemistry, thanks to the sponsorship of Goldberger Textile Works. For a while, he served as rector. István Hazay's (1901–1995) specialty was geodesy—land-survey. In addition to his pedagogical and research activities at the Budapest Technical University, he contributed to solving practical problems on the surveying needs of Hungary. He was a member of HAS.

Bust of Zoltán Csűrös in building "CH."   Bust of István Hazay on the campus.

László Kozma (1902–1983) was an electrical engineer, a graduate of the Brno Technical University (Czechoslovakia at the time; today, Czech Republic). He worked in Belgium, but in 1942, with the German invasion, he lost his job, because he was Jewish. He returned to Hungary and in 1944, he was deported to a concentration camp. He survived, but was severely ill when he returned to Hungary. He worked in industry and taught at the Budapest Technical University until 1949, when the communist dictatorship arrested him on false charges, tried, and sentenced him. He spent six years in prison; in 1955, he was rehabilitated and soon continued his teaching. In 1958, he built the first digital and programmable computer in Hungary. He worked out plans for modernization of the Hungarian telephone system. He reformed the training of electrical engineers, and initiated the specialization of electronic technology. He was a member of HAS.

István Barta (1910–1978) was an electrical engineer. He studied in Germany and graduated from Karlsruhe Technical University. He worked in industry and was a professor and chair of the department of communications engineering at the Budapest Technical University. His research was on radio and television techniques, as well as acoustics. He was instrumental in the development of the Hungarian electronics industry.

The building of informatics, building "I" on the new campus of BUTE on Magyar Tudósok Boulevard, District XI.

Busts of László Kozma and István Barta in building "I."

Árpád Macskásy (1904–1977) was a mechanical engineer and a graduate of the Budapest Technical University. In 1950, he founded the department of building engineering and chaired it for a quarter century. Alfréd Bardon (1904–1986) was a versatile architect and art historian; he chaired the department of drawing. In 1960, he had to take early retirement, but he continued his creative activities and wrote books on architectural history.

Bust of Árpád Macskásy in building "D" (by Sándor Ágh Fábián, 2004) and memorial plaque (on the wall of Macskásy's home, at 15 Bertalan Lajos Street, District XI (by Jolán Humenyánszky).

László Palotás (1905–1993) was an architect, in charge of the department of building materials. He was much involved in designing bridges and other structures. Beside his bust on the campus, he is remembered by a plaque on the house where he lived at 4 Edömér Street, District XI. László Heller (1907–1980) was a mechanical engineer and he and his associate, László Forgó, invented a cooling technique utilizing air condensation. Power stations still use this method. He founded the department of energy studies.

Pál K. Kovács (1907–1989) was a mechanical engineer and a member of HAS who worked in electrical engineering, both in industry and at the Budapest

Left: Bust of Alfréd Bardon on the campus (by Imre Keresztfalvi, 2006).
Right: Bust of László Palotás (by Gábor Szabó, 2005) on the campus.

Left: Bust of László Heller (László Marton, 2007) in building "D." *Right*: Plaque of Pál K. Kovács at 2 Vám Street, District I.

Technical University. From 1970, he continued his career in the United States and elsewhere, but returned home toward the end of his life. Frigyes Pogány (1908–1976) was an architect and art historian. He founded the department of urban architecture and art history at the Budapest Technical University. He designed the reconstruction of the Royal Palace in Buda. András Lévai (1908–2003) studied in Budapest, Graz (Austria), and Vienna. He was a specialist in energy economics and represented a holistic approach, taking into account national, economic, and environmental interests. He held high positions at the University, the Academy, and in the Ministry of Heavy Industry.

László Erdey (1910–1970) was a chemist specializing in analytical chemistry. He was a member of HAS. Emil Mosonyi (1910–2009) was a water works

Left: Bust of Frigyes Pogány (by Bálint Józsa) on the campus. *Right*: Bust of András Lévai (by Sándor Ágh Fábián, 2008) in building "D."

Left: László Erdey's bust in building "CH". *Right*: Bust of Emil Mosonyi on the campus.

architect of international renown and a member of HAS. Because of his involvement in the 1956 Revolution, he was dismissed from the Budapest Technical University and his membership of HAS was suspended, but he was rehabilitated in 1991. He spent two decades as professor of water works at Karlsruhe Technical University.

László Gillemot (1912–1977) was a mechanical engineer, involved in modern metallurgy. He was rector and a member of HAS. József Gruber (1915–1972) was

Left: Busts of László Gillemot and *right*: József Gruber (by Imre Varga, 2002) on the campus.

an aerodynamicist and rector. He set a new course for ventilation construction and production in Hungary.

Károly Simonyi (1916–2001) was a nuclear physicist and an electrical engineer, as well as a science historian. His career had a good start, but it came to a halt due to his involvement in the 1956 Revolution. Only after the political changes, in 1993, was he elected to membership of HAS. Simonyi founded

Károly Simonyi's relief (by Gábor Veres, 2011) on the left is on the wall of the interior of building "Q" of the new campus of the Budapest Technical University (on the right), on Magyar Tudósok Bulevard, District XI.

Left: Memorial plaque and relief honoring Károly Simonyi at the physics campus of the Hungarian Academy of Sciences in Csillebérc, District XII.
Right: The first particle accelerator in Hungary designed by Károly Simonyi. It stands as a memorial in the hallway of the physics department of Eötvös University.

Photo by and courtesy of János Megyeri, Budapest.

Frédéric Joliot-Curie's bust (by László Marton) is on the physics campus of the Hungarian Academy of Sciences in Csillebérc, District XII.

Photo by and courtesy of János Megyeri, Budapest.

research in nuclear physics in Hungary. For many years, he worked on his monumental *Cultural History of Physics*, translated into many foreign languages.[3]

Frédéric Joliot-Curie (Jean Frédéric Joliot, 1900–1958) was a French physicist who changed his surname after he and Irène Curie married. In 1935, the two shared the Nobel Prize in Chemistry for the discovery of artificial radioactivity. Originally, Joliot Curie's bust stood on a square named after him. This honored not only the scientist but also the pro-Soviet politician. After the political changes, the square was renamed and the bust was removed and it disappeared. The plaster mould still existed and the bust was subsequently recast and unveiled in its current location.

Gyula Strommer (1920–1995) was a mechanical engineer, who chaired the department of projective geometry and taught for 53 years at the Budapest Technical University. Frigyes Csáki (1921–1977) was a mechanical engineer and specialist of control and information studies and of automation. He held high functions at the Hungarian Academy of Sciences and at the Budapest Technical University.

Left: Gyula Strommer (by József Kampfl, 2005), mechanical engineer; *right*: Frigyes Csáki (by Iván Paulikovics, 2007); both busts are on the campus of the Technical University.

Statue in memory of the fallen heroes. Jenő Bory's work was originally unveiled in 1930 to honor the heroes of World War I. It was recast and unveiled again in 2000, to symbolize the fallen heroes of World Wars I and II and the 1956 Revolution.

The Budapest Technical University was one of the cradles of the 1956 Revolution and there are various remembrances of this event and the involvement of students and professors on the campus.

Two memorial plaques are in the Aula of building "K." One commemorates Károly Jubal, engineer and instructor, who participated in the War for Independence, 1848–1849, and afterwards, was a member of the national

The assembly hall, the Aula, in the central building "K" of the Budapest University of Technology and Economics.

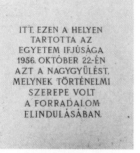

ITT, EZEN A HELYEN
TARTOTTA AZ
EGYETEM IFJÚSÁGA
1956. OKTÓBER 22-ÉN
AZT A NAGYGYŰLÉST,
MELYNEK TÖRTÉNELMI
SZEREPE VOLT
A FORRADALOM
ELINDULÁSÁBAN.

The text on the floor plaque reads: "On October 22, 1956, the university youth held the meeting here that had a historic role in initiating the Revolution."

1956 memorial as part of the fence facing the River Danube (by sculptor Róbert Csíkszentmihályi and architect Róbert Schilling, 2006).

Memorials honoring the 1956 Revolution on the grounds of the campus.

resistance. He was executed on February 28, 1853. The other plaque was erected in memory of those who fell in the 1956 Revolution: Márton Albrecht, József Bartók, János Danner, József Havas, Ferenc Lajos, Géza Péch, János Rössler, József Szilágyi, and Vilmos Vedrál.

There are numerous reliefs and other artistic expressions on the campus that are related to science and engineering. Below we show two from the decorations of building "R": one at the entrance to the building and the other from the back of the building.

Above left: In memory of Károly Jubal (1817–1853); *right*: in memory of those who fell in the 1956 Revolution.

Reliefs on building "R" of the campus (by Csúcs Ferenc, 1954).

So far we focused our attention on the Budapest University of Technology and Economics; now we cover more engineers and innovators who have commemorations in Budapest. They have been graduates of Budapest schools or may have come from far-away places. We start with one of the internationally best-known Budapest graduates and his magic cube.

Ernő Rubik (1944–) accomplished the almost impossible under the socialist regime; he followed through a private initiative, became successful, and thrived. In 1967, he graduated from the Faculty of Architecture of the Budapest Technical University. Following post-graduate studies and working as an architect, in 1975, the College of Applied Art (Iparművészeti Főiskola) appointed him professor. By then he had invented his most famous product, the Magic Cube, also known as Rubik's Cube. It is a three-dimensional puzzle with each of its six faces consisting of nine squares; a colored sticker covers each square, using all together six solid colors: white, red, blue, orange, green, and yellow. The faces may have all their nine squares each display the same color or they may be scrambled. In the case of a scrambled state, the task is to turn around the faces—they can turn around independently—until each face shows a single color.

The tour de force was to enable the faces to turn independently of each other. Rubik initially designed his cube as a pedagogical tool to demonstrate independent motion of components in a device, but soon he recognized its potential as a puzzle. It took years to mass-produce and market, but he persevered. The toy has acquired international fame, it has sold well over 350 million units, and has created a great circle of fans. There have been international competitions for who can be the fastest to unscramble the cube.

Representation of Ernő Rubik's Magic Cube (designed by Ernő Rubik, 2010) in the Graphisoft Industrial Park, off Záhony Street, District III, and on a Hungarian stamp.

View of the Roman city Aquincum from Szentendrei Avenue, District III.

On the way from downtown Budapest toward the Graphisoft Industrial Park, on the right on Szentendrei Avenue, there are the magnificent remains of the ancient Roman city Aquincum. Its exhibitions contain artifacts of Roman culture and science, including the medical arts.

The Graphisoft Park has several pieces of art; there is a Steve Jobs (1955–2011) statue claimed to be the first of this great innovator and entrepreneur. He co-founded Apple, Inc., and pioneered the personal computer revolution.

The plaque at Steve Jobs's statue quotes him: "The only way to do great work is to love what you do." His statue by Ernő Tóth (2011) stands in the Graphisoft Industrial Park off Záhony Street, District III. Jobs is depicted as making a product presentation with his left arm gesticulating and his right hand holding an iPhone.

There are statues of two world-renowned British inventors standing high and decorating the façade of the largest railway station in Budapest. They commemorate George Stephenson and James Watt at Keleti pályaudvar (Keleti, in short), or Eastern Railway Station, Baross Gábor Square, District VII. The station began operations in 1884. In 1892, it acquired its present name, not because of its geographic location, but from the name of the railway company.[4] Architect Gyula Rochlitz designed the building and János Feketeházy and Mór Than designed the hall for the railway tracks. In 1884, Keleti was one of the most modern European railway stations and one of the first using electricity for illumination.

Main façade of Keleti Railway Station, Baross Gábor Square, District VIII.

The four figures above the entrance symbolize a miner for heavy industry, a woman spinner for light industry, a woman harvester for agriculture, and a tradesman for commerce. The two men in the sculptural group on the top symbolize the gods of fire and water and the woman in between symbolizes steam. All of the statues described above and the Stephenson and Watt statues, described below, are not the originals. Some time in the mid-1930s the statues disappeared—it is supposed that they had been made of low-quality stone and had fallen apart. The statues visible today were all made after World War II, relying on original drawings and photographs.

On the northern side of the station, there is a beautiful hall called the Lotz Hall after Károly Lotz, the painter of most of its frescos. They are about the railways and their interaction with the rest of economic life.

Sculptures symbolizing (*from left to right*): heavy industry, light industry, agriculture, and commerce on the main façade of Keleti Railway Station.

The sculpture on the top of Keleti Railway Station. The two men symbolize the gods of fire and water and the woman in the center symbolizes steam.

The statues of George Stephenson (by Ferenc Vasadi, 1884) in the northern statue niche and James Watt (by Alajos Stróbl, 1884) in the southern statue niche on the main façade of Keleti Railway Station.

James Watt (1698–1782) was born in Greenock, Scotland, into a family of merchants. He did not show anything particularly remarkable in his studies until he took up mathematics, at which he excelled. He studied in Glasgow and London, then, returned to Scotland, and worked as a reputed engineer. In 1763, he was asked to repair a steam engine, which Thomas Newcomen had developed. Not only did Watt repair the engine, but he introduced changes that made it perform four times more efficiently than Newcomen's original engine. Watt then co-founded a company, patented his new engine, and mass-produced it. His steam engine became a driving force in the industrial revolution.

George Stephenson (1781–1848) was an English engineer who built the world's first public railway line to use steam locomotives. He was born in Wylam, Northumberland, near Newcastle upon Tyne; the region was famous for coal mining. His parents could not read or write and he did not learn these skills until he was 18 years old when he started attending evening classes. Stephenson proved gifted in the mechanical arts; he repaired various machines and at the age of 23 he became an engineman in a coalmine. When he heard about other people developing a locomotive, he designed his own, which became immensely successful as he continued to improve it. He developed the first ever railway, fully independent of animal power.

There is a statue of Gábor Baross (1848–1892) on Baross Gábor Square, in front of the railway station. Baross was a politician dedicated to modernizing commerce and transportation, especially the railway system and navigation; he served as minister of transportation and minister of commerce.

———— ◆ ————

Statue of Gábor Baross (by Antal Szécsi, 1898) on Baross Gábor Square, District VIII.

Three views of the Museum of Transportation at Városligeti Boulevard, District XIV. *Top*: the entrance to the museum. There is a bust of István Széchenyi at the entrance. *Bottom left*: Ádám Clark's bust and a view of the museum. *Bottom right*: a row of busts (*from left to right*) of Dávid Schwartz, Theodore von Kármán, Lajos Martin (partly hidden), Pál Vásárhelyi, János Csonka, and Kálmán Kandó.

The Museum of Transportation opened in 1899 but its history began in 1896 when there was a huge millennial exhibition in the City Park honoring Hungary's thousand-year history. There were makeshift exhibition pavilions, among them one demonstrating transportation in Hungary. The collection was so rich that the organizers did not want to see it dispersed after the festivities. Hence, the Museum of Transportation was established. The building suffered heavy damage in World War II, and only in 1966 did the rebuilt museum reopen.

Three of the busts around the museum have already figured previously (von Kármán, Jedlik, and Vásárhelyi). The others we mention below in order of the year of birth of the scientist/engineer.

Ádám Clark (1811–1866) was a Scottish engineer who came to Hungary and settled there, hence the Hungarian spelling of his first name. He worked on the railway system of the Habsburg Empire and as an architect in Buda and Pest. His namesake, but no relation, William T. Clark (1783–1852), charged him with supervising the building of the Széchenyi Chain Bridge. W. T. Clark was an English civil engineer and a Fellow of the Royal Society. He designed suspension bridges, among them, the Széchenyi Chain Bridge, in 1839. It was the first

Left: Ádám Clark's bust (by Pál Borics) in front of the Museum of Transportation. *Right*: W. T. Clark's memorial plaque at the Pest bridgehead of the Chain Bridge.

The Széchenyi Chain Bridge.

The statue of zero (by Miklós Borsos), the reference point from which to measure distances in Hungary, Clark Ádám Square, District I.

permanent bridge over the Danube, between Buda and Pest, which, from 1873, joined to form the Hungarian capital Budapest. The Széchenyi Chain Bridge opened in 1849. A plaque marks the location where Ádám Clark's home once stood, next to the Museum of Medical History, at Ybl Miklós Square, District I. It states that Ádám Clark built (rather than designed) the Chain Bridge and the Tunnel.

The square named after Ádám Clark is at the Buda bridgehead of the Széchenyi Chain Bridge and there is a tunnel as though a continuation of the bridge. An eternal joke for children and visitors is that when it rains, to keep the bridge dry, it is moved into the tunnel. The retreating Germans blew up the bridge in 1944 toward the end of World War II. After the war, the bridge was reconstructed in its original splendor.

The statue of a zero on Clark Ádám Square is of magnificent simplicity. It signifies the reference point from which distances on Hungary's most important thoroughfares (Nos. 1 to 7) are measured (with the exception of No. 8, which is measured from Székesfehérvár).

We continue with the busts around the Museum of Transportation. Lajos Martin (1827–1897) was a mathematician and innovator and a member of HAS. He was a trained military engineer and was interested in the theory of flying and rocketry. His fields included ballistics and hydraulics, and he designed screw propellers for ships. At one time he was the Rector of the University of Kolozsvár (today, Cluj-Napoca in Romania). In his inauguration speech he envisioned a new system of transportation, independent of the railways, including airplanes, and the necessity for international agreements on various issues relating to traffic.

Busts of Lajos Martin (by Nándor Záhorzik, 1968) and Dávid Schwarz (by István János Nagy) in front of the Museum of Transportation.

Dávid Schwarz (1850–1897) was a self-trained designer of airships. There has been conflicting data on the fate of his inventions. Actual machines were built according to his plans in St. Petersburg and in Berlin. He did not live to see his constructions fly, but he is credited with laying considerable groundwork for the airships that did eventually fly.

János Csonka (1852–1939) was another self-made man and a great inventor. He and Donát Bánki jointly invented the carburetor for the internal combustion engine. He was born into the family of a smith in Szeged in southern Hungary. János Csonka was interested in how machines worked, in all things that used mechanics for their operation, and in learning everything that might help him to acquire further knowledge about machines and that included foreign languages. He received the best vocational training in his father's workshop, and at the age

Busts of János Csonka at the Museum of Transportation (*left*, by Károly Vasas) and on the campus of the BUTE (*right*, by Sándor Ágh Fábián, 1989). There is a copy of this bust on Fehérvári Avenue, District XI.

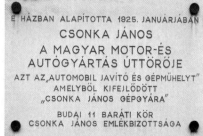

There is a private Csonka János Museum in the house where his machine factory developed; at 31 Bartók Béla Avenue, District XI.

of nineteen, he embarked on a study trip. He travelled to Vienna, Germany, and Switzerland, and broadened his knowledge not only in mechanics, but also in general culture. He spent longer periods in Paris and in British cities.

The possibility of a job that was tailor-made for Csonka lured him back home in 1877. The Budapest Technical University was looking for someone to be in charge of its training workshop. Csonka's ability to speak foreign languages was an extra attraction for the university. It also welcomed Csonka's offer to employ skilled workers at his own expense to help him with his innovations. Csonka perfected machines that had been imported, and he constructed improved versions. His designs included motors driven by gaseous fuel alone and motors that today we would call hybrid, as they were operated either by gas or by liquid fuel.

Csonka was instrumental in forging creative interactions between university and industry, in particular the Ganz Works, whose director general, András Mechwart, was also an inventor in his own right. Donát Bánki, a young engineer at the Ganz Works, and Csonka became friends and they famously augmented each other. Their joint inventions improved the Ganz motors and people referred to them as Bánki–Csonka motors. Their most important invention was the carburetor for regulating the fuel flow into the motor and this was exhibited in 1900, at the Paris world exhibition. Their professional paths diverged when Bánki was appointed head of a department and Csonka became busy with setting up a large new laboratory, but their friendship continued. Csonka never graduated from university, but in 1924, the Budapest Technical University recognized his services and the Hungarian Chamber of Engineers authorized him to use the title, mechanical engineer.

The entrance to the Museum of Electro-techniques at 21 Kazinczy Ferenc Street, District VII, and a display in the staircase.

Statue of Electricity (by Béla Ohmann, 1932) on the facade of the building at 22–24 Honvéd Street, District V.

The Museum of Electro-techniques not only houses memorabilia and a rich library, but also conducts hands-on exercises for classes of schoolchildren.

———•◆•———

Kálmán Kandó (1869–1931) was born in Pest into a family of nobility. He studied mechanical engineering at the Budapest Technical University. He worked in France and Italy before, in 1894, returning to Budapest to the Ganz Works at András Mechwart's invitation. Kandó devoted his entire oeuvre to the electrification of the railway system. He again spent a long period in Italy, working for the Ganz Works, helping to build Europe's first fully electrified main railway line, the Valtellino line, in northern Italy near the Swiss border. He addressed his tasks using a scientific

Kálmán Kandó's bust (by János Sóvári) at the Museum of Transportation (*left*) and a plaque (*right*) honoring Kálmán Kandó at 6 Kandó Kálmán Street, District II. The plaque mentions, among other items, that the international fame of the Ganz Electric Works rested upon Kandó's introduction of the manufacturing of induction motors and that he was a pioneer of the electrification of railways and the designer of the electric locomotive, named after him.

Memorial plaque in the arrival hall of Keleti Railway Station honoring Kálmán Kandó and celebrating the fiftieth anniversary of the inception of modern electric trains, in 1932, in Hungary.

approach, and he introduced many significant innovations, and was elected a member of HAS.

Kornél Zelovich (1869–1935) graduated from the Budapest Technical University and worked for the Hungarian State Railways before he was appointed professor at his alma mater. He was a member of HAS and for a brief period served as rector. His principal achievements were in developing railway lines and railway bridges.

———•◆•———

Bust of Kornél Zelovich (by János Horváth) at the Museum of Transportation.

The Foundry Museum is at 20 Bem Street, where Ábrahám Ganz's Foundry used to be. Ábrahám Ganz (1814–1867) established his foundry in 1858 according to

Bust of Ábrahám Ganz ironworker and manufacturer (by István Cserenyei-Kaltenbach) and the plaque identifying the Ganz Foundry at 20 Bem Street, District II.

The Foundry Museum and the plaque for Gyula Kiszely, the founder of the museum.

Two views of the row of busts in the garden of the Foundry Museum.

his own design. Historian of technology, Gyula Kiszely (1911–1997), founded the museum and he was its first director.

In addition to busts of Ábrahám Ganz and Áron Gábor (below), there are nine busts of engineers and foundry workers in the garden. They are shown here in order of increasing year of birth.

In addition to the bust commemorating artillery officer and gun foundry-man Áron Gábor (1814–1849) in the garden of the Foundry Museum, there

Henrik Fazola (1730–1779), foundryman, by János Andrássy Kurta.

Tivadar Rombauer (1803–1855), metallurgical engineer, by János Andrássy Kurta.

Antal Péch (1822–1895), mining engineer, by Béla Markup.

Ferenc Técsey (1838–1906), metallurgical engineer, by János Andrássy Kurta.

Antal Kerpely (1839–1907), metallurgical engineer, by Béla Markup.

Lajos Katona (1866–1933), metallurgical engineer, by Eszter Balás.

Samu Zorkóczy (1869–1934), metallurgical engineer, by János Csiszér.

László Jakóbi (1897–1957), foundryman, by János Andrássy Kurta.

is another memorial dedicated to him at the corner of Gábor Áron Street and Szilágyi Erzsébet Alley, also in District II. There is a stone plinth with a relief of Áron Gábor and the remains of an exploding cannon ball on the top of the plinth. Gábor participated in the War for Independence of 1848–1849 and was blown apart by cannon fire.

András Mechwart's (1834–1907) name has figured above in the description of careers of inventors related to locomotives and motors. He himself was also an innovator. He was born in Germany and arrived in Pest in 1859, at the invitation of Ábrahám Ganz, to work in his factory. When Ganz died, Mechwart became general manager and directed the Ganz Works between

Gyula Nándori (1927–2005), metallurgical engineer.

Left: The bust of Áron Gábor (1814–1849), gun foudryman, by Eszter Balás, in the garden of the Foundry Museum; *right*: memorial dedicated to Áron Gábor (by Sándor Kiss, 1980) at the corner of Gábor Áron Street and Szilágyi Erzsébet Alley, also in District II.

Mechwart's bust (by András Kocsis, 1965) stands in Mechwart Garden, off Keleti Károly Street, District II. The figure of foundryman on the right (by Alajos Stróbl, 1913) was originally a side figure to András Mechwart's statue, which was destroyed in World War II. This statue now stands on the grounds of the Foundry Museum.

1874 and 1899. The initially 60-member company, by the time Mechwart ended his tenure, employed 6000 people.

———•◆•———

Two engineers have been commemorated for their achievements in various aspects of water economics. Vilmos Zsigmondy (1821–1888), a member of

HAS, was the uncle of Nobel laureate Richard Zsigmondy (see Chapter 1). Vilmos trained as a mining engineer and pioneered research on geothermal energy. He is best known for his achievements in drilling artesian wells. There is a memorial plaque on the house where Zsigmondy lived at 95 Dob Street, District VII.

Lajos Bertalan (1838–1901) was a water engineer and, as chief engineer, he directed the flood prevention efforts for the Tisza River.

János Irinyi (1817–1895) was a chemist who had invented a new match even before his career had taken off. He replaced chlorate of potassium with lead dioxide, which made the match, which had previously been noisy and unsafe, quiet

Vilmos Zsigmondy's bust (by Antal Szécsi) stands at Kós Károly Walkway in City Park, District XIV.

Lajos Bertalan's commemorating plaque (by András Lapis, 2005) at 28 Bertalan Lajos Street, District XI, refers to his engineering the protection against flooding from the Tisza River.

Two statues of János Irinyi. *Left*: The statue on Mikszáth Kálmán Square, on the corner building at Szentkirályi and Reviczky Streets, District VIII. This was the location of the match factory, which Irinyi founded. *Right*: The statue by Barna Búza (1965) stands in the small garden at 21–23 Lágymányosi Street, District XI.

and safe. The income from the sale of his invention financed his studies. In the 1848–1849 War for Independence, he directed the production of gunpowder and gun foundry. After the defeat and having served his prison term, he devoted the rest of his life to chemical research.

Tivadar Puskás (1844–1893) traveled widely in Europe and the United States. He learned about the telegraph and the telephone and they fired up his imagination. He invented the idea of the central telephone exchange for connecting several people talking. In America, he worked with Edison. Back in Europe, in Paris, Puskás built the first telephone exchange on the continent in 1879. Two years later, he built the first telephone exchange in Budapest, which was the fourth in the world. He was concerned with the broad issues of transmitting the human voice and music.

Ferenc Puskás (1848–1884) was Tivadar's younger brother. He served as an army officer, but resigned his commission at his brother's suggestion. Ferenc became a pioneer in setting up telephone exchanges when there were still only a few dozen telephones in Budapest. After an initial lukewarm reception for the idea of the telephone, this business soon blossomed. He exhausted himself and died young.

A 1931 relief decorates the entrance to the building, which used to house the "József" telephone center. This imposing building opened in 1917 and was the first building in Budapest to use reinforced concrete. It is now abandoned and waiting for a new use.

Relief at 17–19 Horváth Mihály Square, Distict VIII, honors Tivadar and Ferenc Puskás. The image on the right highlights their portraits on the relief.

Bust of Tivadar Puskás in a small park at 55 Krisztina Boulevard (the headquarters of Hungarian Telekom), District I (by Ferenc Tischler, 2006).

Statue of Tivadar Puskás (by Gyöngyi Szathmáry, 2009) at Szombathelyi Square, District XI, commemorates the 125th anniversary of the first telephone center.

Tivadar Puskás's bust in the Puskás Tivadar Technical High School of Telecommunication at 22 Gyáli Avenue, District IX.

Endre Magyari (1900–1968) graduated from the Budapest Technical University as a mechanical engineer. In 1922, he started to work at the Experimental Station of the Hungarian Post Office. He was a pioneer of broadcasting in Hungary and directed the first broadcast from a makeshift station, a wagon used for transporting furniture. This wagon now stands in the courtyard of the Puskás Tivadar Technical High School of Telecommunication, next to Magyari's bust.

———— • ◆ • ————

Endre Magyari's bust in the courtyard of the Puskás Tivadar Technical High School of Telecommunication at 22 Gyáli Avenue, District IX. The wagon standing next to Magyari's bust was the first broadcasting station in Hungary.

The inventor of movable print, Johannes Gutenberg's memorial is on a building on Gutenberg Square. It was built between 1905 and 1907 in the secessionist style—Art Nouveau—and once belonged to the Benevolent Society of Book Printers and Type Casters, hence the connection with Gutenberg. In the middle of this square, is József Fodor's memorial statue; he was a pioneer of the public health service in Hungary (see Chapter 4).

Relief in memory of Johannes Gutenberg, the inventor of movable print on the wall of the building, 4 Gutenberg Square, District VIII.

László József Bíró (1899–1985) is not a household name, but the "biro" is, as it symbolizes the modern ballpoint pen. Budapest journalist Bíró noticed that the ink from his fountain pen did not dry quickly on paper and smudged easily. In contrast, the ink used for printing newspapers dried quickly and was smudge-free. Bíró tried the printing ink in his pen, but it was too viscous and would not flow. He and his chemist brother, György, constructed a new tip for a pen that contained a small ball, freely turning around in its socket, picking up just the right amount of ink from the adjacent cartridge for writing on paper. In 1938, the brothers patented their pen. In 1943, the Jewish Bíró brothers fled Hungary, and

László József Bíró's memorial plaque on the wall of his home at 12 Cimbalom Street, District II. The Hungarian postage stamp honoring him was issued in 1996.

moved to Argentina. There, Bíró was known as Ladislao José Biro. He sold the patent to the owner of the Bic company, and this is why Bíró's ballpoint pen has acquired the alternative names of biro and bic. The Day for Inventors in Argentina is celebrated annually on Bíró's birthday, September 29.

Notes

1 József Németh, *From the BME (Budapest Technical University) to World Renown* (Budapest: Budapest University of Technology and Economics, 2007), p. 36.

2 C.G. Bernhard, "Presentation Speech for the Nobel Prize in Physiology or Medicine 1961." In *Nobel Lectures, Physiology or Medicine 1942–1962* (Singapore: World Scientific, 1999), pp. 719–721; actual quote, p. 721.

3 In English: Károly Simonyi, *A Cultural History of Physics* (Boca Raton, FL: CRC Press/A. K. Peters, 2012; David Kramer, translator). Simonyi's son, Charles Simonyi, directed the English edition of the book. Charles Simonyi has lived in the United States since he was a teenager. He is a software engineer, entrepreneur, and philanthropist.

4 This is also how the other major Budapest railway station, the Western Railway Station (Nyugati pu.), which opened in 1877, received its name, from the company operating the station.

7
László Rátz
SCHOOLS OF CREATIVITY

Left: Early photo of László Rátz. *Right*: Plaque in the stairway at the Lutheran Gimnázium, 17-21 Városligeti Alley, District VII.
Courtesy of Zsolt Szabó, Lutheran Gimnázium, Budapest.

The high school mathematics teacher László Rátz (1863–1930) has become a legend. One of his former high school pupils, the Nobel laureate physicist Eugene P. Wigner, never failed to praise him, even during his two-minute Nobel Banquet Speech in Stockholm, on December 10, 1963. We chose László Rátz to symbolize this chapter about the Budapest high schools, the *gimnáziums*, and to discuss a sample few in detail.

At the time of Rátz's activities, the Hungarian high school was at its best. Someone characterized the excellence of the *fin de siècle* Hungarian high school by observing that the Hungarian high school is still quite good although it has deteriorated over more than a hundred years. Wigner said that "Budapest was filled with fine high schools," referring to the early twentieth century.¹ In 2001, the Nobel Museum in Stockholm organized an exhibition for the centennial celebrations of the Nobel Prize. The exhibition had two main sections: one was devoted to individual creativity and the other to creative milieus. The organizers selected the famous Budapest gimnáziums to represent one of fifteen subsections on creative milieus.

A pioneer in Hungarian pedagogy, János Apáczai Csere (1625–1659), could be another choice to symbolize this chapter. He was a philosophical and pedagogical writer; a forerunner of the Enlightment. He studied and worked in Holland, but spent the last few years of his short life in Transylvania, building up educational institutions.² A practicing gimnázium of Eötvös University carries his name, along with other educational institutions, streets, and a prestigious award in education.

Much of the fame of the Budapest high school is due to two small groups of their alumni. One group consists of the seven Nobel laureates that had graduated from five high schools. The other is the so-called Martians—the five scientists who, in addition to their discoveries, became important contributors to the defense of the United States and the free world. There are several variants of the story about the origin of the Martians label. According to one, a group of scientists in the Manhattan Project talked about the conspicuous presence of gifted Hungarians in the project and someone suggested that they spoke Hungarian as a disguise, but had really come from the planet Mars. There were many other future luminaries training at the same time.

Left: Statue of János Apáczai Csere (by István Kiss, 1966) in the park at the meeting point of Irinyi József, Bogdánfy, and Budafoki Streets, District XI. *Right*: Bust (by Mihály Mészáros, 1959) at the entrance of the Apáczai Csere János Practicing Gimnázium of Eötvös Loránd University at 4–6 Papnövelde Street, District V.

The photograph of the teaching staff of the Lutheran Gimnázium is scary. These teachers were typical of the *fin de siècle* high schools. To be a gimnázium teacher created an aura and provided a good salary as well. They were all male, whereas today, with much lower prestige and salary, a large proportion of gimnázium teachers are female. In the early twentieth century, it was not rare for these teachers to engage in research; a few would even become members of the Hungarian Academy of Sciences.

The teachers of the Lutheran Gimnázium at around the turn of the nineteenth to the twentieth century, sitting in front of the Sundial-fresco.
Courtesy of Zsolt Szabó, Lutheran Gimnázium, Budapest.

The renovated Sundial-fresco on the Lutheran Gimnázium in 2012.
Courtesy of Zsolt Szabó, Lutheran Gimnázium, Budapest.

The gimnázium had a regimented system. Classes started with recitation and the teacher could call upon any student at any time to be tested; this was mostly done orally. This created tension; some teachers terrorized the class. Decades later, our own experience was not that different. We do not think that the later success of pupils should be ascribed to such a nerve-wrecking atmosphere.

The future Nobel laureates and the future "Martians" were only a small sample of the pupils. The former group (see also Chapter 1) included the engineer–physicist Dennis Gabor, the pharmacist turned economist John Harsanyi, the chemists George de Hevesy and George A. Olah, the writer Imre Kertész, the biochemist Albert Szent-Györgyi, and the chemical engineer turned theoretical physicist "Martian" Eugene P. Wigner. The other "Martians" were the aerodynamicist Theodore von Kármán, the physicist Leo Szilard, the chemical engineer turned mathematician John von Neumann, and the physicist Edward Teller.

The five "Martians" were all born into upper-middle-class Jewish families within a small area of Budapest. They attended excellent high schools before emigrating to Germany, and later, to the United States. One often hears that they all went to the same high school, but this is a myth. In fact, the five Martians attended three different high schools and if we add the high schools attended by the future Nobel laureates, the number of high schools rises to seven. We restrict our coverage to mentioning these seven, but stress that there were other high schools also providing outstanding education at the time and producing graduates who also became luminaries through their creative activities.

When the Martians immigrated to the United States, they had traveled from Germany, and sometimes one reads that they were German. This impression became even stronger due to the fact that two of them have "von" in their names, which implies German nobility. The fathers of von Kármán and von Neumann acquired hereditary nobility in the Austro-Hungarian Monarchy. Their names did not originally contain "von." In Hungarian, nobility is easy to recognize by a forename before the surname. This forename refers to the location of the nobleman's estate. Thus, for example, von Neumann's full name in Hungarian was margittai Neumann János where "margittai" refers to the place Margitta where the Neumanns' estate was supposed to be. However, when they left for Germany, they understood that in Germany, the Hungarian way of indicating nobility would not work, and they adopted the simpler German way of expressing it. One might say that they had no authorization to use "von" in their names, but nobody worried about such details. It is amusing that such highly intelligent men as von Kármán and von Neumann cared for such a distinction.

All five Martians grew up in assimilated families, although the degree of assimilation varied. Eventually, four of the five converted; Edward Teller was the only exception. It is safe to say that their conversion was due to expediency; it was supposed to provide protection from anti-Semitic persecution, although it did not. In 1919, after World War I and the ensuing revolutions, Szilard wanted to re-enter the Budapest Technical University where he had been a student before being conscripted.

However, nationalist students prevented him from entering and beat him up. Szilard produced the certificate of his conversion, but nobody was interested.

John von Neumann waited for conversion until after his father died, in 1929. There is no evidence that it mattered much to him, until his tragic terminal illness of a brain tumor, when—according to gossip—he sought solace in the Catholic religion, alas, to no avail. Edward Teller and his bride went for their marriage ceremony to a Calvinist church in Budapest to follow her religion—her Jewish family had previously converted—but this made no difference to Teller's world view, in which religion had no role to play.

The label Martian underwent some evolution during the 1980s in Hungary. When the communist political system was gradually giving way to relaxation,

The Calvinist church on Városligeti Alley, where the Tellers married in 1934.

Below: The newly-weds Edward and Mici Teller.
Courtesy of the late George Marx.

High-school graduation pictures of Theodore von Kármán, Leo Szilard, Eugene P. Wigner, John von Neumann, and Edward Teller.
Courtesy of the Archives of the Hungarian National Museum, Budapest (TvK), the late Ferenc Szabadváry (JvN), and the late George Marx (LSz, EPW, and ET).

these great scientists were becoming heroes in their home country as well. The physicist George Marx did much to popularize them. He kept adding other expatriates who had become successful in the West, and the Martians label became an umbrella term, extending to economists, movie directors, and even chess players. The Martian label also served as a euphemism for Jewish origin. Here, we restrict the meaning of this label to its original group of five scientists, previously listed.

The five Martians knew each other, but they did not form a group; they interacted among themselves, and today we would call it networking. Their years of birth span three decades; yet their paths show great similarities, as though they ran along parallel lines. While high school education excelled in Hungary at the time of their youth, university education was not remarkable.

Lutheran Gimnázium

Miklós Vermes (1905–1990) was a high school teacher of mathematics, physics, and chemistry. He graduated from the Lutheran high school in Sopron, western Hungary, the same school from which two other notable teachers had graduated before him, László Rátz and Sándor Mikola. Vermes attended the University of Budapest and earned his degree in experimental physics. He began teaching at the Lutheran Gimnázium where he taught until 1952, at which time the communist authorities closed the school. Vermes rescued a good deal of the physical apparatus for teaching, which he used in the Jedlik Ányos Gimnázium in Csepel (the southernmost district of Budapest). He wrote well-received textbooks in physics, and was active in organizing competitions for pupils in physics. He received the highest state awards for his pedagogical work.

The Lutheran Gimnázium some time in the past (*left*) and in 2012 (*right*). It is the Budapest Fasori Evangélikus Gimnázium, at 17–21 Városligeti Alley, District VII.
Courtesy of Zsolt Szabó, Lutheran Gimnázium, Budapest.

Plaque at the entrance to the school, commemorating former pupils Eugene P. Wigner, John von Neumann, and John Harsanyi, and former teachers, Ödön Hittrich, Sándor Mikola, László Rátz, and János Renner.

Plaque and bust of Miklós Vermes (by Tamás E. Soltra, 2012) in front of the school.

Busts of János Balogh (by Károly Bartha, 2004) and John Harsanyi (by Ilona Barth, née Mezőfi Mózer, 2004).

There are four busts of former pupils in the school. There is one of János Balogh on the ground floor and of three émigré scientists on the first floor, John Harsanyi, Eugene P. Wigner, and John von Neumann.

János Balogh (1913–2002), future zoologist and ecologist and a member of HAS, attended this school, then, he graduated from Budapest University as a doctor in zoology and as a high school teacher of biology and geography. Eventually he became chair and professor at the department of zoological taxonomy and ecology of Budapest University. He participated in over thirty expeditions to the tropics organized by UNESCO. He was dedicated to the cause of the protection of nature.

Both Eugene P. Wigner and John von Neumann attended the Lutheran high school as Jewish pupils (the Wigners converted during his last year of high school studies). The tuition of the Jewish pupils was multiple times that of the Lutheran students.[3] However, parents considered it worthwhile if they could afford it, and the school had many Jewish pupils. László Rátz taught mathematics and when he recognized von Neumann's talent, he gave him private lessons, free of charge. Rátz also appreciated the gifted Wigner and gave him books to study. Wigner later noted that he did not match von Neumann's excellence and their teacher had made the correct distinction. Today, there is a national-level teacher's award named after Rátz.

Eugene P. Wigner began his university studies in chemical engineering at the Budapest Technical University. From 1921, he continued his studies in Berlin in chemistry at the Technical University. The instruction was quite traditional; in the inorganic chemistry course, instead of electrons orbiting the nucleus, he learned about materials and properties. This knowledge proved invaluable later when he

Busts of Eugene P. Wigner (1993) and John von Neumann (1993). Both by Ilona Barth, née Mezőfi Mózer.

worked on nuclear reactors. Wigner was a conscientious chemistry major. Being aware that his father directed a tannery in Budapest and that he expected his son to join him there as a chemical engineer, Wigner took special courses that he thought might be useful in the tannery. Nevertheless, his fascination was only for physics.

Wigner was proud of his knowledge of chemistry and materials and was versed in the various processes of the tannery. Later he utilized his knowledge in chemical engineering within the Manhattan Project and, after the war, in directing Clinton Laboratories, from which today's Oak Ridge National Laboratory developed. Wigner has been often referred to as the world's first nuclear engineer.

Wigner recognized when in Berlin that the University and its weekly colloquia provided a unique opportunity for him to immerse himself in his beloved physics. He met Einstein through Szilard and visited him often in his home. They talked about physics as well as about social and political questions. In the meantime, Wigner continued his chemistry studies and acquired his Diploma in Chemistry (equivalent to a Master's degree) and his doctorate. The famous polymer chemist Herman F. Mark was his mentor for his Diploma work, the project being the symmetries of the rhombic sulfur crystal. This is how he initially embarked on symmetry studies, which then became a life motif for him. Although Wigner was a student of the Technical University, he carried out his research at the Kaiser Wilhelm Institute in Berlin-Dahlem, where Mark had a position under Michael Polanyi. Polanyi then became Wigner's mentor for his doctoral studies. After his doctorate, Wigner returned to Budapest for about a year and worked as chemical engineer at the Mautner Tannery.

EBBEN A HÁZBAN SZÜLETETT
WIGNER JENŐ
1902 – 1995
NOBEL – DÍJAS FIZIKUS,
A FASORI EVANGÉLIKUS GIMNÁZIUM
EGYKORI DIÁKJA,
AZ EÖTVÖS LORÁND TUDOMÁNYEGYETEM
DÍSZDOKTORA,
AZ EÖTVÖS LORÁND FIZIKAI TÁRSULAT
TISZTELETBELI TAGJA.
MAGYARSÁGÁT MINDIG VÁLLALTA.
AZ ATOMREAKTOROK KIFEJLESZTÉSÉBEN
ÉS AZ ATOMFIZIKÁBAN AZ EMBERISÉG
SZÁMÁRA MARADANDÓT ALKOTOTT.

Eugene P. Wigner's birthplace at 76 Király Street, District VI. The plaque on it mentions, among other items, that Wigner was a Nobel laureate physicist, he attended the Lutheran Gimnázium, he was an honorary doctor of Eötvös Loránd University and an honorary member of the Eötvös Loránd Physical Society, and that he always considered himself to be Hungarian. It also mentions that Wigner created eternal values for humankind in developing atomic reactors and atomic physics.

Eugene P. Wigner (standing at the back) with associates of the Mautner Tannery.
Courtesy of the late Martha Wigner Upton.

Soon though he was invited back to Berlin and his career in theoretical physics took off. Wigner accepted a part-time job in the United States even before the Nazis came to power in Germany. He helped Szilard in getting Einstein to prepare and sign his famous letter to President Roosevelt in the summer of 1939. Einstein's letter warned the American leadership of the possibility of a German atomic bomb.

Wigner was active in war-related research. Szilard called him the conscience of the Manhattan Project. Wigner maintained contact with the engineers who were to build the nuclear reactors, which provided the fuel for the atomic bombs. The design work had to go ahead barely in front of the engineering activities, since everything was new; they were all pioneers and had to move along untrodden paths.

Wigner was an arch-conservative who viewed everything coming from the East with suspicion. Yet he retained a genuine interest in everything Hungarian and was able to recite poems he had learned in the gimnázium impeccably throughout his life. In 1976, he visited Budapest for the first time after the war and this was followed by two more visits in the late 1970s and the 1980s.

Eugene P. Wigner and one of the authors in 1969 at the University of Texas at Austin.

Wigner's arrival in 1976 at the Eastern Railway Station. Wigner carries the bouquet; István Kovács is on his right; George Marx is in between the two, behind them; and Imre Tarján is on Wigner's left, somewhat behind. Kovács, Marx, and Tarján were physics professors.

Courtesy of the late George Marx.

Eugene P. Wigner in 1976 at the entrance to the "red hedgehog house" in the Castle District, at 3 Hess András Square, District I, with the decoration above the entrance.

Courtesy of the late George Marx.

Wigner in his Princeton office with a portrait of László Rátz.
Courtesy of the late George Marx.

Wigner with Michael Polanyi and his son (the future Nobel laureate) John in 1934 in Manchester, England.
Courtesy of John C. Polanyi, Toronto.

Whenever people asked Wigner about the geniuses who had come out of Hungary, he always responded that there was only one genius and that was John von Neumann. According to the Nobel laureate German–American physicist Hans Bethe, von Neumann represented a higher-level organism, a mutation of ordinary human beings.

Below: John von Neumann's birthplace, 26 Báthori Street, District V, the corner of Báthori Street and Bajcsy-Zsilinszky Avenue. *Right*: The plaque on the house refers to von Neumann as a world-famous Hungarian scientist and pioneer of computational technology.

EBBEN A HÁZBAN SZÜLETETT

NEUMANN JÁNOS

1903—1957

A VILÁGHÍRÜ MAGYAR TUDÓS,

A SZÁMÍTÁSTECHNIKA ÚTTÖRÖJE

AZ EMLÉKTÁBLÁT A NEUMANN JÁNOS
SZÁMÍTÓGÉP – TUDOMÁNYI TÁRSASÁG 1987-BEN
ÁLLÍTOTTA, MAJD 2004-BEN ÚJRA FELAVATTA.

Neumann's family later moved to 16 Arany János Street, District V.

John von Neumann's graduating class of 1921. Von Neumann is the second from the left in the second row from the top.

Courtesy of Zsolt Szabó, Lutheran Gimnázium, Budapest.

John von Neumann with his daughter Marina.

Courtesy of Marina von Neumann Whitman, Ann Arbor, MI.

The former Neumann villa on Eötvös Avenue, District XII with a plaque at its entrance. The plaque notes that the villa belonged to the Neumann family from the 1920s through World War II and that the family used to spend part of their summers in the villa.

Von Neumann was born into a well-to-do family. He made utmost use of their rich home library. In addition to their home in downtown Budapest, the von Neumanns maintained a villa in a beautiful area of the Buda Mountains.

Von Neumann manifested special gifts in mathematics from early childhood. At one point, his mathematics teacher, László Rátz, determined that he could not teach him anything more, and asked Professor József Kürschák of the Technical University for advice. At Kürschák's suggestion, Gábor Szegő became von Neumann's tutor. Szegő eventually ended up in the United States, at Stanford University, as a world-renowned mathematician. What he did was not really tutoring; mentor and pupil got together weekly, and amid sipping of tea, they discussed mathematical problems of mutual interest. Von Neumann had already begun to publish scientific papers, even during his high school years.

In spite of his obvious mathematical talents, von Neumann initially did not embark on a career in his favorite subject. At his parents' insistence, he studied chemical engineering in Berlin and Zurich. In parallel with his chemistry studies, however, he earned a doctorate in mathematics from Budapest University. He settled in Germany and had an academic appointment. He sensed as early as the mid-1920s that the conditions in Europe would be worsening and that there might be another war coming. He considered the growing economic prowess of the United States to be inevitable and that the country would strive to achieve a leadership position in science.

Von Neumann proved to be tremendously productive in the 1920s, but when an invitation came from Princeton, he was ready to move. The appointment was

Left: John von Neumann's bust (by Gyöngyi Szathmáry) at the Infopark, District XI, on the joint campus of the Budapest University of Technology and Economics and Eötvös Loránd University. *Right*: Bust at the Puskás Tivadar High School of Telecommunications, 22 Gyáli Avenue, District IX.

at first on a part time basis and both he and Wigner alternated between Princeton and Germany, but the situation changed with the Nazis coming to power, and von Neumann and Wigner settled in the United States for good. Von Neumann felt comfortable in his new home country; its informal style of life agreed with him (in contrast with Wigner who did not feel comfortable with the informality). Von Neumann became a professor of the recently established Institute for Advanced Study. He dealt with a plethora of problems, and his interest and talent extended to the most diverse fields of the applications of mathematics. When the US war efforts involved his participation, he did not let any specific project bog him down; rather, he became a sort of roving ambassador and moved from project to project. After the war, he focused his attention on developing the modern computer.

Mintagimnázium

Today, the school's official name is Trefort Ágoston Practicing Gimnázium of Eötvös Loránd University. It is at 8 Trefort Street, District VIII. Of the five Martians, von Kármán and Teller went to the secular *Mintagimnázium* (Model high school). Von Kármán's father, Maurice, was an educational expert. At his initiation, the Model developed into a high school of practice teaching for teachers training at Budapest University. The school opened in 1872, and Maurice Kármán directed it until 1896. Earlier, Maurice Kármán had served as mentor to one of the members of Franz Joseph I's royal family. When the emperor-king offered him a high reward for his services, Maurice declined saying that he would prefer

Left: The Minta, now Trefort, Gimnázium, at 8 Trefort Street, District VIII. *Right*: Co-discoverer of the double helix James D. Watson and his wife, Elizabeth (on his right), in Magdolna Hargittai's company in 2000, in front of the plaque honoring Mór (Maurice) Kármán.

something that he could hand down to his children. According to family legend, this is how he acquired heritable nobility.

There used to be two principal directions for gimnáziums: the "classical" and the "modern." The classical placed emphasis on literature, history, and classical languages; the modern stressed the sciences and modern languages. Although the Minta followed the classical curriculum, it was not in any respect behind the "modern" school. Through their instruction, the classical subject, the dead Latin language, came to life in the classroom. They taught mathematics in conjunction

Left: Theodore von Kármán's birthpace at 22 Szentkirályi Street, District VIII. *Right*: Memorial plaque on the house referring to von Kármán as the president of the International Astronautic Academy, world-renowned researcher of the technological sciences, and pioneer of aviation science and rocket technology.

with its practical uses; for example, the pupils prepared graphs, determined rates of change, and thus acquired an understanding of the basics of calculus, which was not yet in the curriculum. Tódor (Theodore) attended the Mintagimnázium between 1891 and 1899.

Theodore von Kármán visiting his father's grave (plot 5A-18-1) in 1962 at the Kozma Street Jewish Cemetery, District X. The inscription says, "Does not die that lived for the sake of so many." ("Nem halt meg az ki oly sokaknak élt.")

The photo is courtesy of Roger Malina, Richardson, Texas.

Left: Edward Teller's boyhood home was in the building on the left, 3 Szalay Street, District V. *Right*: One of the authors (IH) and US diplomat, Deputy Chief of Mission Jeffrey Levine, in front of Teller's memorial plaque on the day it was unveiled, January 15, 2008. The plaque carries an inscription in English and in Hungarian. (The building of 3 Szalay Street extends between Honvéd Street on the east side and Vajkay Street on the west side. The Teller apartment faced Vajkay Street, but the memorial plaque is on the wall of the building on Honvéd Street. The reason is that a clerk of a second-hand bookstore on Honvéd Street initiated the erection of the plaque.)

The main entrance to the Hungarian Parliament building with the two lions.

Edward Teller attended this school one generation later, between 1917 and 1925. He had a mixed experience at this school; first he had an inspiring maths teacher; later a dull one. He complained especially about his physics teacher, who set Teller back several years although physics was his favorite subject. During his high school years, Teller's family lived at 3 Szalay Street, in a large apartment house, and occupied a spacious third floor home. On the centenary of Teller's birth, a memorial plaque was unveiled on the building.

As a child, Teller often played at Kossuth Lajos Square, a few hundred yards from his home. Even decades later, he remembered the two lions standing guard in front of the main entrance to the Parliament building. When Nelson Rockefeller asked Teller if he would like anything from Budapest, Teller mentioned the images of these lions.

Following his graduation from the Minta, Teller became a student of the Budapest Technical University (more about this in Chapter 6). In 1926, he left Hungary

Left: Edward Teller on the bank of the Danube on the Buda side with the Franz Joseph Bridge. *Right*: The same location today; the bridge is now called Szabadság Bridge.
Courtesy of the late George Marx.

for Germany, and in 1933, he moved on, first to Copenhagen, then to London, finally, in 1935, to the United States. Until 1936, Teller visited Budapest regularly, kept up with colleagues in physics, and welcomed foreign visitors to his hometown.

Beyond the two "Martians," other future luminaries of science went to this school, among them, two future British economists, Tamás (Thomas) Balogh, Baron Balogh (1905–1985), and Miklós (Nicholas) Káldor, Baron Kaldor (1908–1986), and two Polanyis, Karl and Michael. Karl Polanyi (1886–1964) was a world-renowned economist–historian and Michael Polányi (1891–1976) was a medical doctor turned physical chemist turned philosopher. Their brother Adolf (1883–1966) and their sister Laura (1882–1957) also became noted contributors to world culture. Laura attended the Lutheran Gimnázium, but only as a private student with special permission. These high schools were all-boys-schools at the time.

The four Polányi siblings in 1948, Adolf, Karl, Laura, and Michael.
Courtesy of the Hungarian National Museum, Budapest.

Michael Polanyi achieved fundamental results in chemistry, uncovering the mechanism of the adsorption of gases on solid surfaces, the structure of fibers, and an understanding of how chemical reactions occur. He experienced considerable frustration when he found it difficult to get his teachings accepted. He had broad interests and background and gradually he moved toward philosophy, especially epistemology. His main work was a book entitled *Personal Knowledge*, first published in 1958; it has become a classic. It is about the personal element of what we know.

Left: The Polanyi memorial plaque at 2 Andrássy Avenue, District VI, with Egyptian-American chemistry Nobel laureate Ahmed Zewail, some time in the 1980s. *Right*: The current (2012) plaque. Unknown vandals have repeatedly destroyed the Polanyi memorial plaques.

Courtesy of Ahmed Zewail, Pasadena, California.

Further famous pupils of the Minta included the future Oxford low-temperature physicist Nicholas Kurti (1908–1998), and the 2005 Abel Prize winner Peter Lax (1926–) of New York University. The successor school to the Minta still operates in its original building, and participates in teacher training.[4]

Nicholas Kurti in 1994 in London and Peter D. Lax in 2007 at New York University.

Main Modern Gimnázium, District VI

Leo Szilard went to the *Főreálgimnázium* (Main Modern Gimnázium) of District VI. This was a splinter institution from the *Főreálgimnázium* of District V, on Markó Street (see below). Szilard's school was in a beautiful building, completed in 1898, and was state-of-the-art in its equipment and outstanding in the teaching staff. The building still stands, part of it housing a research institute, but much of it sadly abandoned.

Left: Leo Szilard's birthplace at 50 Bajza Street, District VI. There is a memorial plaque on this house, both in English and Hungarian: "Leo Szilard, brilliant physicist, was born here on 11 February 1898. He was one of the founders of information theory, designer of the first atomic reactor, and worked tirelessly to prevent the proliferation of nuclear weapons and to promote the peaceful use of nuclear energy." *Right*: The Vidor Villa— Szilard's home—at 33 Városligeti Alley, District VII. A plaque on the building explains that Leo Szilard was a physicist, pioneer of atomic research, and one of the initiators of the peaceful utilization of atomic energy.

The building of the former Main Modern Gimnázium, District VI. *On the left*, the main body along Rippl Rónai Street; *on the right*, the wing extending into Szondi Street.

Leo Szilard's graduating class in 1916. Szilard is third from the right in the front row.
Courtesy of the late George Marx.

Leo Szilard after World War II in Chicago.
Courtesy of the US Department of Energy, Washington, DC.

At the time of his high school studies, between 1908 and 1916, Szilard's family lived at 33 Városligeti Alley in the Vidor Villa, named after Szilard's architect uncle who designed the building for the extended family. For Szilard, the Main Gimnázium was quite close, but the Lutheran Gimnázium would have been yet closer. The Main Gimnázium was nondenominational, which was probably an attraction for Szilard.

Leo Szilard was the most colorful of the "Martians." He did not live by the rules of convention and held it that if he had to choose between being tactless and being untruthful, he preferred to be tactless. He was always ahead of everybody else by a few steps. This made him a source of valuable advice for many. However, some disliked that his mind raced so fast that he would come to conclusions based on other people's data before those producing the data could come to them on their own.

Szilard left Hungary at the very end of 1919, became a student in Berlin, and completed his studies there including a PhD degree. When the Nazis came to power, he left for England. In fall 1933, Szilard came to the idea of the nuclear chain reaction. He considered that if he could find an element that neutrons could split, and which would *emit two* neutrons for every *one absorbed* neutron, such an element, in a sufficiently large amount, could sustain a nuclear chain reaction, and could even explode. Szilard thus arrived at two concepts. One was the nuclear chain reaction—it was really a branched chain reaction—and the other was the critical mass. Szilard assigned his patent on the nuclear chain reaction to the British Admiralty in order to keep it safe from the Germans.

When in 1938, Otto Hahn and Fritz Strassmann discovered nuclear fission of uranium, experimentally, and Lise Meitner and Otto Frisch interpreted it as such, Szilard knew that uranium was *the* element. He saw the danger that Germany might develop the atomic bomb and enlisted Albert Einstein to sign a letter of warning to the American president, Franklin D. Roosevelt. This letter started the events that eventually led to the Manhattan Project and the first atomic bombs.

Szilard was ten years old when he first read Imre Madách's magnum opus, *The Tragedy of Man*. The piece "amazed him and later in life shaped his mental and moral speculations."[5] There is some similarity between Goethe's *Faust* and Madách's *Tragedy*, in that in both, the Devil tempts Man and his faith in God. In fifteen scenes,

Statue of Imre Madách (by Tibor Vilt, 1973) at the rose garden on Margaret Island.

the *Tragedy* follows the history of humankind, and predicts the future. Szilard drew lessons from Madách's work that never left him. One of them was that "it is not necessary to succeed in order to persevere" and another was that there is always "a narrow margin of hope."[6] This lent Szilard optimism in all his activities.

Markó Street Gimnázium, District V—Berzsenyi Gimnázium

The Berzsenyi Dániel Gimnázium operated for almost a hundred years, from 1858. It was the first state high school, and from 1941, it united two former independent high schools. One of them came to life at 35 Markó Street and the other at 18–20 Markó Street. The one at 35 Markó Street was the first state school, yet denominational—Catholic. Instruction began in 1858 in a different location, and for the first two years, in German; they then switched to Hungarian. This building, at 35 Markó, served the school from 1876 onwards.

In 1883, the school acquired the building at 18–20 Markó Street, which was designed by Alajos Hauszmann. In 1952, when the Berzsenyi Gimnázium had to vacate its building, it left behind its excellent library and its well-endowed apparatus and laboratories, and moved to another location. We do not know the exact reason why the school had to leave, but the Internet home page of today's Berzsenyi Gimnázium gives a clue: "After the academic year of 1951–52 following a high-level [communist] party decision, Berzsenyi was moved to District XIII as a slightly disguised punishment. It had to leave its building, equipment, furniture and library behind."[7] What the punishment was for is difficult to discern. Our guess is that the school represented an independent spirit that in 1952 was anathema to the rigid communist regime.

Left: The former gimnázium at 35 Markó Street (today, it is no longer a high school).
Right: The building at 18–20 Markó Street, District V, used to be the Berzsenyi Gimnázium; today it is the Xantus János Bilingual Teacher Training Secondary School. *<http://www.xantuski.hu> (downloaded November 18, 2013).*

During the war years and afterwards, the school could not be free from the impact of the politics surrounding it, yet indications are that it withstood the tests of these periods better than most. Following the anti-Jewish legislation in 1939, the *numerus clausus* kind of restrictions previously applicable to university education were extended to high schools. Only a token number of Jewish pupils were admitted, except for a few Budapest gimnáziums in which entire Jewish classes were organized. Berzsenyi was one of them and thus helped its Jewish pupils to continue a school life with a semblance of normalcy until about April 1944.[8]

The Berzsenyi Gimnázium produced stellar contributors to both the Hungarian and American societies. They included the president of the short-lived democratic Hungarian Republic of 1918–1919, Mihály Károlyi (1875–1955). He used

Statue of former president Mihály Károlyi (by Imre Varga, 1975).
Photo, at the original location of the statue, by and courtesy of József Varga, Budapest. Currently the statue is in Siófok.

Statue of the martyr poet Miklós Radnóti (by Imre Varga, 1969) at 11 Nagymező Street, District VI.

to have a beautiful statue near the Parliament building, but in 2012, the reigning administration removed it (ostensibly, to redesign the square where the statue stood, but Károlyi had long been anathema to the right and extreme right). There are plaques for the martyr poet Miklós Radnóti (1909–1944), the writer Frigyes Karinthy (see, in Chapter 4), and the philosopher Bernát Alexander (1850–1927), on the wall of the school.

Two busts decorate the entrance to the school, one on each side; one is of the school's current eponym, János Xantus (1825–1894), and the other, the painter István Szőnyi (1894–1960). Xantus was an ethnographer and geographer; he

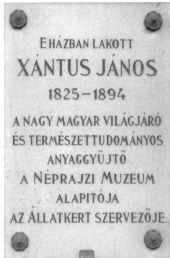

Left: Bust of János Xantus. *Right*: The memorial plaque on the house where Xantus lived, at 44 Damjanich Street, District VII, mentions that he was a great Hungarian traveler and collector of scientific material; he founded the Museum of Ethnography and organized the Budapest Zoo.

Bust of István Szönyi (by Imre Varga, 1977) in front of the Xantus János high school.

participated in the Revolution and War for Independence, 1848–1849. During the ensuing period, he spent about a dozen years in the United States before returning home. He was a great collector of ethnographic materials and samples of flora and fauna in the United States and brought his tremendous collection back to Hungary. He was a member of HAS. Szőnyi was a painter of great renown and a humanist, whose family members helped save persecuted Jews and others during the war.

The renowned future professor of the Budapest Technical University, Donát Bánki, studied here. Bánki later became Theodore von Kármán's mentor at the Budapest Technical University (Chapter 6).

Alfréd Hajós (1878–1955) was also a graduate of the school; Arnold Guttman was his original name, but he changed it and chose one connected with water (hajós means sailor or seaman). At the age of 13, he decided to become a good swimmer after his father drowned in the Danube. In 1896, he was 18 when he won two gold medals at the Athens Olympics for 100 meters and 1200 meters free style swimming; they were Hungary's first Olympic medals. Later, as a graduate of the Budapest Technical University, he became a noted architect, and designed the swimming stadium on Margaret Island. His tombstone at the Kozma Street Jewish Cemetery displays the five-ring symbol of the Olympic Games.

Another former pupil, Marcell Grossmann (1878–1936), was a mathematician, a friend and fellow student of Albert Einstein, and professor of mathematics at what today is the Swiss Federal Institute of Technology (ETH) in Zurich. The future Hungarian-American financier George Soros (1930–) was another pupil of Berzsenyi Gimnázium. He has built an empire of wealth and founded the Central

Plaque on the wall of John Kemeny's boyhood home, 38 Bajcsy-Zsilinszky Avenue, District V.

European University (CEU) in Budapest. The late Tom Lantos (1928–2008), long-time Californian Congressman, studied also at the Berzsenyi. He was the founder of the Congressional Human Rights Caucus, which was renamed the Tom Lantos Human Rights Commission in 2008. He served as chair of the Foreign Relations Committee of the US House of Representatives.

Another pupil was the future president of Dartmouth College (Hanover, NH), mathematician John G. Kemeny (1926–1992), who earned fame first for his contributions to computer languages and later for serving as the chief examiner of the Challenger disaster. There was merely a single block between the boyhood homes of John Kemeny and John von Neumann (see, above).

A future Rector of Budapest Technical University, Gusztáv Rados (see Chapter 6), was another of the school's pupils.

George Polya (1887–1985) was a Hungarian-American mathematician. He studied in Budapest and Vienna, and in 1912, he was Lipót Fejér's doctoral student. In the same year, he left Hungary, moved to Zurich, and he worked as professor of mathematics at the ETH between 1928 and 1940. From 1940, he lived in the United States and was professor of mathematics at Stanford University. He wrote books that are now classics in the field. He was an honorary member of HAS.

Georg Klein (1925–) is an internationally renowned immunologist in Sweden. For many years he served on the Nobel Assembly of the Karolinska Institute in Stockholm. He has received numerous awards and prizes, and he is probably one of the most decorated scientists of our time. (He is in the photograph together with Imre Kertész, depicted in Chapter 1.)

László Polgár (1946–) is the author of well-known chess books and the father and chess teacher of the chess celebrities— former prodigies—Zsuzsa (Susan, 1969–), Zsófia (Sofia, 1974–), and Judit (1976–).

There was a future Nobel laureate among the pupils of this school, Dennis Gabor. His path ran parallel to the Martians. From Hungary he moved to Germany, and then to Great Britain. Gabor trained as a physicist–engineer and acquired skills and knowledge that would guarantee his employment in any non-racist country. In England he had research and development jobs in industrial laboratories, and then, professorial appointments. He invented holography, and received the Nobel Prize in Physics in 1971.

The mathematician George Polya.
Courtesy of Gerald L. Alexanderson, Santa Clara, CA.

The house where the Gábors lived when he was born, at 25 Rippl Rónai Street, District VI; the house is just across the street from Leo Szilard's high school. There is a plaque at the entrance to the building. It remembers Dénes Gábor, the creator of holography; Nobel laureate scientist; humanist thinker.

Left: Plaque on the wall of Xantus János high school (former Berzsenyi Gimnázium) honoring Nobel laureate Dennis Gabor. *Right*: There is another plaque to the right from the entrance, at a later home of the Gabor family, at 30 Falk Miksa Street, District V.

Piarist Gimnázium

The Catholic high school of the Piarist Order was the alma mater of two future chemistry Nobel laureates, George de Hevesy (or, Georg von Hevesy, depending on the language he was using, see Chapters 1 and 3) and George A. Olah (see Chapters 1 and 6). The Piarist Order has taught in Pest since 1717, and the present location has housed a Piarist school since 1762, and has operated as a high school since 1848. Their new building was completed in 1917, but the school had to vacate this in 1953 and move to another location (Mikszáth Kálmán Square, District VIII). In 2011, they returned to their original, beautifully reconstructed building at Pesti Barnabás Street, District V.

The two principal corpuses of the recently renovated Gimnázium of the Piarist Order at the Pest Bridgehead of Erzsébet Bridge. The inscription beneath the cupoled tower on the corpus of the school: COLLEGIUM ORDINIS SCHOLARUM PIARUM ERECTUM A D MDCCCCXV.

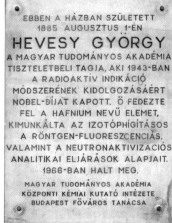

George de Hevesy's birthplace at 3 Akadémia Street, District V; just across the street from the Hungarian Academy of Sciences. The plaque on its wall enumerates Hevesy's achievements.

Georg von Hevesy at around 1907 in Freiburg, Germany.

Courtesy of László Kovács; László Kovács, Hevesy György 1885–1966 (Szombathely: Studia Physica Savariensia VI, 2000).

Hevesy attended the Piarist Gimnázium as a Jewish student between 1895 and 1903; his family converted at around the time of his graduation or soon after. In their tolerance, the Piarists built on their previous liberal traditions.

At the time of the Holocaust, the school had its Jewish pupils remove the yellow star from their clothing while in the building. Oláh provides this laconic statement: "I do not want to relive here in any detail some of my very difficult, even horrifying, experiences of this period, hiding out the last months of the war in Budapest."[9] In the spring of 1945, he completed his studies at the Piarist Gimnázium, which, regardless of the immediate post-war conditions, instituted a demanding final examination.

The Olah Family lived at 13–15 Hajós Street, District VI, when George A. Olah was born, just across the street from the Opera House.

The photograph shows Olah as a high school student.

Courtesy of G. A. Olah, Los Angeles.

Lónyai Street Gimnázium—today, Szent-Györgyi Albert General School and Gimnázium

Another future Nobel laureate, Albert Szent-Györgyi, went to yet another high school, in Lónyai Street; today the school bears his name (there is more about Szent-Györgyi in Chapter 1).

Left: Albert Szent-Györgyi's image as a relief in the Aula of the Center of Theoretical Medicine of Semmelweis University, Szentágothai János Square, District IX. *Center and right*: Busts of Albert Szent-Györgyi in elementary schools named after him: *Center*: in the garden of the school at 20 Csömöri Street, District XVI (by Róza Pató, 1993); *right*: in the hallway of the school at Szérüskert Street, District III (by József Kampfl, 1994).

The Szent-Györgyi Albert Gimnázium today in Lónyai Street, District IX.

The young Albert Szent-Györgyi, not long after graduation.
Courtesy of Andrew Szent-Györgyi, Waltham, MA.

Madách Gimnázium

This school bears the name of the great writer Imre Madách whose most famous drama *The Tragedy of Man* has been on theater programs for decades. It has had an impact on many people, among them, Leo Szilard (see above). The school

Madách Imre Gimnázium at 5 Barcsay Street, District VII, with two frescos on its façade.

was founded in 1881 as the District VII State Gimnázium, and in 1920, it became the Hungarian Royal Madách Imre State Gimnázium. Today, its name is simply Madách Imre Gimnázium. The school's website lists only a few scientists as celebrity graduates, but many more names in literature, film and theater, music, the arts, the media, sport, and politics. The most noteworthy is the Nobel laureate in literature, Imre Kertész (1929–), whose award made the Madách the fifth Budapest gimnázium from which future Nobel laureates had graduated. György Konrád (1933–), another writer, is also a famous graduate, and so is the mathematician Pál Turán (1910–1976).

Pál Turán with his family. His wife, Vera T. Sós, is also a famous mathematician.
Courtesy of Vera T. Sós, Budapest.

Andy Grove (1936–), yet another Madách graduate, is missing from the listings on the school website. He was born as András Gróf, a name he changed when he moved to the United States after the 1956 Revolution. After the Madách, he attended Eötvös University as a chemistry major, and continued his studies at City College of New York, from which he graduated in 1960 with a bachelor's degree in chemical engineering. Grove did his graduate studies at the University of California at Berkeley and he received his PhD degree there in 1963. He was a pioneering contributor to both the semiconductor revolution and to the computer revolution. He served as the CEO of Intel Corporation. Opinions are divided as to whether he was greater as an engineer or a businessman. *TIME* magazine presented him as "Man of the Year" on its cover on December 29, 1997.

György Konrád with István Hargittai in 2005.

Andy Grove with Eszter Hargittai.
Courtesy of Eszter Hargittai, Evanston, IL.

Notes

1 I. Hargittai, *The Martians of Science: Five Physicists Who Changed the Twentieth Century* (New York: Oxford University Press, 2006; 2008), p. 3.

2 Transylvania was a semi-independent Hungarian principality, a vassal state of the Turkish Empire. Much of Hungary was under Turkish occupation and the remaining part—the Kingdom of Hungary—was under Habsburg domination.

3 We have accurate information only from later years: the ratios of tuition for Lutheran, Calvinist, Catholic, and Jewish pupils were 1, 2, 3, and 5, respectively, in the academic year 1927/1928, according to archival data from Mr. Zsolt Szabó of the Lutheran Gimnázium. According to the school's yearbook, published in 1943, the respective ratios were 1, $2\frac{2}{3}$, 4, and $6\frac{1}{2}$ in the academic year 1943/1944.

4 Our daughter, Eszter Hargittai, attended this school between 1989 and 1992 and our son, Balazs Hargittai, attended the Apáczai Csere János Practicing Gimnázium between 1985 and 1988.

5 William Lanouette with Bela Silard, *Genius in the Shadows: A Biography of Leo Szilard—The Man Behind the Bomb* (Chicago: University of Chicago Press, 1992), p. 23.

6 Lanouette, *Genius in the Shadows*, p. 23 and p. 24.

7 The web site of the current Berzsenyi Gimnázium is <http://berzsenyi.hu/index.php?id_a=25>.

8 See, e.g. Janos M. Bak, "1939–47 . . . and Since." *The Hungarian Quarterly*: Vol. 50, No. 195, 2009.

9 George A. Olah, *A Life of Magic Chemistry* (New York: Wiley, 2001).

8

Gedeon Richter

MARTYRS AND SAVIORS

"Shoes on the Bank of the Danube"
(by Can Togay and Gyula Pauer, 2005)
commemorates the Jewish martyrs
murdered by the Hungarian Nazis, the
Arrow-Cross, between 1944 and 1945.

Martyrs

There is a unique memorial a few steps northbound from the Hungarian Academy of Sciences, along the lower embankment of the Danube. Sixty pairs of shoes of cast-iron commemorate the many thousands of Jewish martyrs shot into the river in the fall of 1944 and winter 1944–1945. It was the time of the horror reign of the Hungarian Nazis, the Arrow-Cross, during the last months of World War

II. The monument was unveiled on April 16, 2005. April 16 has become the day of Holocaust remembrance in the schools of Hungary.

The Arrow-Cross seized power on October 15, 1944. By then, under Regent Nicholas Horthy, over 400,000 Jews, mostly from outside of Budapest, had been deported to Auschwitz where the vast majority perished. Then, the Arrow-Cross murdered an estimated 100,000 Jews. They hunted out groups of people and shot them into the Danube, often after they had been tortured. The murders took place all along the river on the Pest side, and the place chosen for the monument was symbolic.

Above: Holocaust Memorial by Imre Varga, in the Wallenberg Park, on Wesselényi Street, District VII, behind the Dohány Street Synagogue. *Left*: James D. Watson in 2000, visiting the memorial.

Gedeon Richter's bust is at the headquarters of the Richter pharmaceutical company, 19–21 Gyömrői Avenue, District X.

Photo by and courtesy of Sándor Görög, Budapest.

Plaque on the building at 21 Katona József Street, District XIII, where Gedeon Richter lived. The plaque, by István Buda, states that on December 30, 1944, Gedeon Richter was taken from this building and murdered.

There are other beautiful general memorials to Holocaust victims in Budapest, but very few dedicated to individuals.

There were scientists among the victims, including the pharmacist manufacturer–businessman Gedeon Richter.[1]

Gedeon Richter (1872–1944) was born into an assimilated Jewish landowner family, in an East Hungarian village Ecséd, near the town of Gyöngyös. Before completion of his high school studies, he started working in a pharmacy in Gyöngyös. It was at the time that new legislation required university training for pharmacists, and he successfully completed the prescribed higher education. After this, he spent two years working in various pharmacies to acquire the qualifications to open his own pharmacy. He soon found his real vocation in the creation of the independent Hungarian pharmaceutical industry. He spent four years in Italy, Germany, France, and England, preparing himself for the task. He was 29 years old when he bought a pharmacy and started his own manufacturing laboratory using the money from the sale of his family's estate. The pharmacy he ran is still there, at the corner of Üllői Avenue and Márton Street in District IX.

There is a drug store at the same location that Gedeon Richter started his operations, at the corner of Üllői Avenue and Márton Street, District IX.

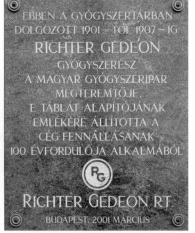

A plaque from 2001 commemorates the centennial of this event.

Richter focused on deficiency diseases, such as avitaminosis. The necessary ingredients for the therapy, called organotherapia, were extracted from animal endocrine glands. He built up a strong research section in his laboratory and kept close contact with the medical profession. By 1902, he had already started publishing an information bulletin and was distributing it among medical doctors, free of charge. The laboratory soon proved inadequate for his goals and he founded a plant for manufacturing his preparations, which extended to plant extracts and synthetic drugs as well. One of his associates, Emil Wolf, soon left him to found another future giant pharmaceutical company, Chinoin.

Bust of Emil Wolf, co-founder of Chinoin, (by Dávid Tóth, 2010) is at 23 Pozsonyi Avenue, District IV (at the corner of István Avenue and Nyár Street).

Emil Wolf (1886–1947) studied in Munich and graduated in 1910 as a chemical engineer. He and György Kereszty (1885–1937) co-founded the chemical factory from which the Chinoin Chemical and Pharmaceutical Company was created in 1913. They produced mainly synthetic drugs in close cooperation with Géza Zemplén, professor of organic chemistry at the Budapest Technical University. In 1944, the Jewish Wolf was deported to a concentration camp in Germany. He survived and returned home, continuing to direct Chinoin, but died shortly after, in 1947.

The Richter chemical factory acquired fame for some long-lasting products, such as kalmopyrin (comparable with aspirin), insulin, and the Glandtrin injection which contained oxytocin and proved efficient in gynecological applications. Insulin was discovered in 1923 and Richter was already manufacturing it by 1926. In the early 1930s, Richter became one of the leading worldwide manufacturers of estrogens. Richter created subsidiaries and expanded internationally. In 1923, the Richter chemical factory became a shareholders' company with Gedeon Richter retaining the majority of shares. By the mid-1930s, Richter products were marketed in 100 countries. By the late 1930s, the Richter Company was second only to the United Incandescent Lamp Company (Tungsram) among Hungarian exporters.

From the late 1930s, increasingly harsh anti-Jewish legislation was hindering normal operations. The company was placed under military control; Gedeon Richter was under attack and had to resign as Chairman of the Board. In 1942, he was stripped of the rest of his functions in the company, and soon he was banned from entering the plant. He offered his services without remuneration, but his offer was rejected. For some time, he was still participating covertly in directing his company through his faithful associates. He rejected suggestions that he should escape while it was still possible. His only concern was how to save the company during these turbulent times.

On December 30, 1944, the Arrow-Cross took him to the embankment of the Danube and murdered him. His corpse was never found. The company has survived; today, it carries Gedeon Richter's name.

————— • ◆ • —————

Lajos Steiner (1871–1944) was a meteorologist and geophysicist. He studied at the University of Budapest and for some time he was Loránd Eötvös's associate. Between 1892 and 1932, he served at the Royal Hungarian Central Institute of Meteorology and Earth Magnetism; and from 1927, as its director. His principal research concerned the theory of geomagnetism. His main achievement was the introduction of the weather forecasting service in Hungary. In 1917, he was elected a corresponding member of HAS. On April 2, 1944, he committed suicide, thus escaping further anti-Jewish persecution.

Nándor Mauthner (1879–1944) was a chemical engineer who studied at the ETH Zurich and graduated in 1902. In 1903, he received his doctorate at the University of Geneva. For a few years, he did research in Emil Fischer's

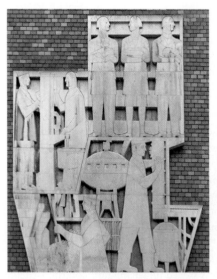

"Biochemical drug manufacturing"—relief on the wall of one of the buildings of the Gedeon Richter Pharmaceutical Company.
Photo by and courtesy of Sándor Görög, Budapest.

Lajos Steiner medal of the Hungarian Meteorological Society by Miklós Borsos.

organic chemistry institute in Berlin. From 1911, he worked at the University of Budapest with one interruption when between 1917 and 1918, he served as a military chemist in Vienna. He received a promotion in May 1919, during the communist dictatorship, and this impacted adversely on his subsequent career. His main field of interest was in sugar chemistry. In 1934, he was elected a corresponding member of HAS—a rare occurrence for a Jewish scientist during the quarter century of Horthy's reign. After the mid-1930s there is no information about Mauthner, except for the rumors that on May 21, 1944, he committed suicide, finding no other way out of anti-Jewish persecution.

Adolf Káldor (1882–1944) was born in Modor (now, in Slovakia). He was the first municipal physician of the town Budafok—today, part of District XXII of Budapest. He was popular among his patients. Together with his family he was deported in late spring–early summer 1944 and perished in Auschwitz, according to the documentation of Yad Vashem, the Israeli Remembrance Authority.

Adolf Káldor's bust (by Erzsébet Schaár, 1969) in front of the medical offices at 2 Duna Street, District XXII.

Imre Bródy (1891–1944) was another martyr scientist. He was born in 1891 in Gyula, in southeastern Hungary. Between 1909 and 1914, he studied mathematics and physics at Budapest University. He began teaching in a high school, and conducted research in theoretical physics at the University. In 1918 he earned his doctorate. His excellence earned him recognition and the university invited him to be assistant professor. Following the period of war and revolutions, the virulent anti-Semitism engulfing the country prevented a university career for Bródy. In 1920, he left for Germany, and joined the great physicist Max Born in Göttingen. Many of the world's best young physicists, among them Werner Heisenberg, congregated around Born. Therefore, Born's words with respect to Bródy's talent carry exceptional weight: "There was the little Hungarian Jew, E. Bródy, perhaps *the most gifted of them all*, who could solve intricate problems . . ."[2] (our italics). Here E. stands for Emerich, the German equivalent of Imre.

Bródy seemed destined for a great career, but after two years he returned to Hungary. He joined the research laboratory of the United Incandescent Lamp Company, known by its trademark, Tungsram. The sagacious director general, Lipót Aschner, developed a research laboratory under the leadership of Ignác Pfeifer. Pfeifer attracted excellent scientists to work there, such as Bródy, Pál Selényi, and others. Even expatriate scientists, such as Michael Polanyi and Dennis Gabor, took up part-time appointments as external advisors.

Bródy demonstrated great versatility when he transformed himself from being a successful theoretician into an innovative technologist. His best-known invention was the krypton lamp, which had great advantages over the previously used argon lamp. The application necessitated new ways of producing the expensive krypton gas, in which Bródy cooperated along with other outstanding scientists, such as Ferenc Kőrösy and Polanyi. The krypton lamp became an international success, and in 1937, the company built a whole new manufacturing plant for the lamp near the town of Ajka.

The intensifying anti-Semitism in Hungary and the increasingly harsh anti-Jewish legislation reached Tungsram as well. The company had to replace both Pfeifer and Aschner. The new man in charge, Zoltán Bay, was both an excellent scientist and a caring human being, but eventually he could not prevent the tragedies that were to come. Bay achieved special status for the company as being important for defense. A number of leading researchers and engineers, including Bródy, received exemption from the slave labor service that Jewish men were subject to. This saved them for some time. However, when Bródy's wife and daughter were deported, he gave up his exempt status. The Nazis arrested him on July 3, 1944. Inmates sighted him at various camps, including Auschwitz–Birkenau. His wife and daughter probably perished soon after their arrival in Auschwitz and by the end of 1944, Bródy too was dead.

Imre Bródy on a photograph (*left*) and a relief (*right*), at the Museum of Electro-techniques, 21 Kazinczy Ferenc Street, District VII.

The physicist and inventor Pál Selényi (1884–1954) survived the slave labor camp, but his two sons were murdered. Selényi graduated as a high school teacher of physics and mathematics and embarked on a career at the University. Following Loránd Eötvös's death he was appointed lecturer in experimental physics. In 1919, he was a member of the leadership of scientific societies and museums.

Right: Pál Selényi and his wife's tombstone in the Kozma Street Jewish Cemetery lists their two martyr sons: György (1915–1944) and Tamás (1923–1944).

Above: Pál Selényi experimenting.
<http://www.muszakiak.hu/feltalalo-fizikus-mernok/magyar-fizikusok/103-selenyi-pal-izzolampa>

After the collapse of the communist dictatorship, he could not hold a state-supported position. He became an associate of Tungsram and excelled with numerous inventions. One of these is considered to be the forerunner of photo-copying. The anti-Jewish legislation forced him into retirement in 1939. He returned from the slave labor camp severely ill, but continued his work; he was elected to the Hungarian Academy of Sciences and taught as a professor at the Budapest Technical University.

Frigyes Fellner (1871–1945) was an internationally renowned economist and statistician. He trained at the law school of the University of Budapest, and in 1897, he earned the qualification of lawyer. He achieved high positions in banking, but when he embarked on a career in academia, he withdrew from direct involvement in finance. He pioneered in Hungary the determination of gross national income; at this time even the Central Office of Statistics did not conduct such analyses. He was a professor at both the University of Budapest and the Budapest Technical University. In 1917, he was elevated to the nobility. He was a member of HAS. In 1927, soon after the Upper House of Parliament formed, Fellner became an alternate member, and in 1938, he became a full member when Pál Teleki resigned his membership (due to Teleki's election to the lower chamber). Fellner was never directly involved in politics.

Fellner's spectacular career ended abruptly. The timing coincided with the introduction of anti-Jewish legislation in 1939.[3] It may be that he had kept his

Frigyes Fellner.

Szenes Fotó: Frigyes Fellner. <http://www.magyarzsido.hu/index.php?option=com_catalogue&view=detail&id=497&Itemid=29> (downloaded 4 April, 2013).

Jewish origins secret or might not have been aware of them. His membership of the Upper House of Parliament was terminated although such membership was for a lifetime. His publishing activities came to a halt in 1939, at which time he had 130 publications. There is no information about his last five years. He was arrested soon after the German invasion of Hungary on March 19, 1944. He was taken to the Mauthausen concentration camp where in early 1945, he starved to death.

Three members of the Hungarian Academy of Sciences mentioned above (Steiner, Mauthner, and Fellner) perished between 1944 and 1945 because of their Jewish roots. We cannot know whether there were more. It seems that there has been no research on the losses of the Hungarian Academy of Sciences with respect to its members or other scientists in the Holocaust. A superficial browsing of the volumes of the members of the Academy gives the impression that hardly any Jewish scientists became members of the Academy in the Horthy era: the period between 1920 and 1944. Mauthner's election to corresponding member was a rare exception. Fellner's election is irrelevant in this respect because his possible Jewish background was not known.

Anti-Jewish bias in elections to the Hungarian Academy of Sciences is difficult to investigate. Data on the ethnicity and religion of the members of the Academy are scarce and elections depend on so many factors that demonstration of bias can hardly be unambiguous. There is though a distinct impression that such a bias existed and was strong. This issue should be investigated to end decades of amnesia concerning the history of the Academy of Sciences between the two world wars. We offer only a single example of one person who was not elected in spite of being recognized as worthy of membership.

In 1934, eight members of the Academy nominated John von Neumann for membership. The nominators were outstanding scientists whose scholarship

John von Neumann.
Courtesy of Marina von Neumann Whitman, Ann Arbor, MI.

enabled them to appreciate von Neumann's achievements. In addition to a descriptive recommendation, they listed von Neumann's 49 publications in support of their nomination. The signatories were Ottó Titusz Bláthy, Gusztáv Rados, Radó Kövesligethy, Károly Tangl, Lipót Fejér, Béla Pogány, István Rybár, and Rudolf Ortvay. Alas, von Neumann was not elected. He was 31 years old at the time and looking back from today's perspective when new members of the Academy are often over 60 years of age, one might think that he may have been too young for this honor. However, this was not the case. Just consider von Neumann's nominators, who, apart from Bláthy, were all elected at a young age; Fejér at 28 (in 1908), Pogány at 31 (1918), Rados (1894) and Rybár (1918) both at 32, and Kövesligethy at 33 (1895). In 1937, von Neumann became a naturalized U.S. citizen and in the same year, he was elected a member of the National Academy of Sciences of the U.S.A.

From 1939, the Horthy regime conscripted Jewish men and from 1941, after Hungary entered the war, used them as auxiliary troops—in reality as a slave labor force—on the Eastern Front and elsewhere. In addition to the hardship and dangers that accompany wartime conditions, they suffered from the humiliation of their situation and the sadistic treatment by their supervising Hungarian army personnel, the so-called skeleton staff. To be sure, there were exceptions, but what was more characteristic was the depiction by the Oscar-winning director István Szabó, based on a true story, in his film *Sunshine*. There, a former Olympic sportsman who had successfully represented Hungary in fencing competitions was tortured to death. Many of the inmates of slave labor camps never returned, and there were among them representatives of Hungarian scientific life.

Memorial to the Jewish slave laborers, 1939–1945, at 2 Bethlen Gábor Square, District VII.

Inscription at the Memorial to the Jewish slave laborers, 1939–1945: "... fegyvertelen álltak az aknamezőkön" (they stood unarmed on the minefields). In reference to Miklós Radnóti, A la recherché ...".... standing unarmed on distant and freezing minefields; ... " [Thomas Orszag-Land (translator), *The Witness: Selected Poems* by Miklós Radnóti. London: Tern Press, 1977, p. 42.] Other translations exist. The inscription applies, among many others, to the father of one of the authors (IH). Dr. Jenő Wilhelm was made to sweep a minefield with bare hands and a mine blew him apart in the fall of 1942; his remains rest in a mass grave in Western Russia.

Future members of HAS that had been in slave labor camps (there might be others missing from this list):

György Ádám (1922–2013), physiologist
Pál Benedek (1921–2009), chemical engineer
Frigyes Csáki (1921–1977), mechanical engineer
Ervin G. Erdős (1922–), external member, pharmacologist
Jenő Ernst (1895–1981), biophysicist
Dávid Rafael Fokos Fuchs (1884–1977), philologist
László Fuchs (1924–), external member, mathematician
Tibor Gallai (1912–1992), mathematician
János Gergely (1925–2008), physician, immunologist
István Hahn (1913–1984), historian
Péter Hanák (1921–1997), historian
Róbert Hoch (1926–1993), economist
Miklós Julesz (1904–1972), physician
László Kalmár (1905–1976), mathematician
László Kardos (1898–1987), historian of literature
Béla Kellner (1904–1975), physician, oncologist
György (Georg) Klein (1925–), honorary member, physician, immunologist
Károly Lempert (1924–), chemist
József Lukács (1922–1987), philosopher
Károly Marót (1885–1963), classical philologist
Gyula Mérei (1911–2002), historian
Zsigmond Pál Pach (1919–2001), historian
Alfréd Rényi (1921–1970), mathematician
László (Ladislas) Robert (1924–), external member, biologist
Pál Selényi (1884–1954), physicist
István Simonovits (1907–1985), physician
Bence Szabolcsi (1899–1973), musicologist
Sándor Szalai (1912–1983), sociologist
Pál Turán (1910–1976), mathematician
György Vajda (1927–), mechanical engineer

Imre Vajda (1900–1969), economist
Tibor Vámos (1926–), electrical engineer
Andor Weltner (1910–1978), jurist
Ervin Wolfram (1923–1985), chemist
László Zsigmond (1907–1992), historian

Future members of HAS that had been in concentration camps (there might be others missing from this list):

Iván Berend T. (1930–), economist historian, Dachau
Ervin G. Erdős (1922–), external member, pharmacologist, Sachsenhausen
János Frühlig (1937–), external member, physician, oncologist, Strasshof
István Hargittai (1941–), chemist, Strasshof
Avram Hershko (1937–), honorary member, physician, biochemist, Strasshof
Miklós Julesz (1904–1972), physician, Buchenwald
József Knoll (1925–), physician, pharmacologist, Auschwitz
László Kozma (1902–1983), electrical engineer, Mauthausen-Gunskirchen
Károly Lempert (1924–), chemist, Mauthausen
Géza Mansfeld (1882–1950), physician, pharmacologist, Auschwitz
Pál Pándi (Kardos) (1926–1987), historian of literature, Laxenburg (Austria)
Rezső (Ruben) Pauncz (1920–), external member, chemist, Strasshof
György Ránki (1930–1988), historian, Auschwitz
Gábor Szabó (1927–1996), biologist, Auschwitz
Andor Weltner (1910–1978), jurist, Buchenwald, Dachau

Memorial to the victims of fascism (by Makris Agamemnon, 1986) at Viza Street, District XIII.

A few names appear in both lists. Some survivors of the slave labor camps were then deported to concentration camps.

Makris Agamemnon (1913–1993) was a Greek sculptor who spent a considerable period of his life and career in Hungary, while a military regime reigned in Greece. One of his famous creations is a monumental memorial to the victims of the Mauthausen Nazi concentration camp in Austria. A copy of this memorial, unveiled in 1986, stands in Budapest on a bank of the Danube. It commemorates "the resistance fighters, deserters, and the persecuted whom the Fascists murdered on the Pest bank of the Danube in the winter of 1944–45." The dedication is a typical expression of the ambivalence of the communist party toward the Hungarian Holocaust. The principal victims murdered on the bank of the Danube were Jews and the murderers were the Hungarian Nazis—the Arrow-Cross, but the János Kádár regime did not consider it "politically correct" to spell this out. The renovated memorial was once again unveiled on April 14, 2010. The occasion was marked by an ecumenical service. In May 2012, anti-Jewish hate slogans were painted on the memorial. The perpetrators were not misled by the euphemism of the inscription on the memorial.

There is a plaque with two sets of names in the hall of the Rényi Institute of Mathematics of the Hungarian Academy of Sciences. One of the sets is "Our Greats" and the other is "They Embarked on the Road of Creating." There is then

Left: Rényi Institute of Mathematics of the Hungarian Academy of Sciences, in the middle on the right, 13–15 Reáltanoda Street, District V. *Right*: Memorial plaque for the mathematician victims of Fascism with a line by the poet Miklós Radnóti, himself a victim: "az új falak tövében felhangzik majd szavam."

Miklós Radnóti, ". . . my words, they will yet ring out by those new walls; . . ." "The Protector" [Thomas Orszag-Land (translator), The Witness: Selected Poems by Miklós Radnóti. London: Tern Press, 1977, p. 35.] Other translations exist.

a third list that is conspicuously missing, because it would be for "those young talents that prepared for their start." The memorial plaque was unveiled in 1976; Pál Turán spoke at the ceremony.

"Our Greats"	**"They Embarked on the Road of Creating."**
Mihály Bauer 1874–1945	Ervin Aczél –1942
Pál Csillag 1893–1944	István Ádám 1925–1944
Géza Grünwald 1913–1943	Ervin Feldheim 1912–1944
Dénes Kőnig 1884–1944	József Krausz –1944
Simon Sidon 1892–1941	Dezső Lázár 1913–1943
Adolf Szücs 1884–1945	Gyula Sándor 1921–1945
	Miklós Schweitzer 1923–1945
	István Valkó 1904–1945
	László Waldapfel 1911–1942

Here, we say a few more words about some of the people on the lists.

Mihály Bauer studied at the Budapest Technical University where Gusztáv Rados and Gyula Kőnig were among his teachers. Bauer started writing mathematics papers at the age of eighteen. In 1918, he received his professorial appointment at the Budapest Technical University. In 1922, Bauer was the first recipient of the newly established Kőnig Gyula Prize of the Eötvös Loránd Mathematical and Physical Society. In contrast, he was kept at a low-level position at the Budapest Technical University, where anti-Semitic students often disturbed his lectures. In around 1936, the university forced him into retirement. In 1944, he was taken to a concentration camp from where he returned, but he died in February 1945.

Dénes Kőnig was the son of Gyula Kőnig (Chapter 6). Dénes studied in Budapest and Göttingen. In 1907, he received his doctoral degree and joined the Budapest Technical University. He was a pioneer of graph theory, and in 1935

Dénes Kőnig and his joint grave with his father, Gyula Kőnig, at the Fiume Avenue National Necropolis (plot 10/1-1-12).
Courtesy of Vera T. Sos, Budapest.

he reached the rank of full professor. In 1936, he published a monograph, *Theorie der endlichen und unendlichen Graphen*. It has remained a fundamental treatise in the field and in 1990 it was published in English translation, *Theory of finite and infinite graphs* (Birkhauser). Following the Nazi occupation of Hungary on March 19, 1944, Dénes Kőnig helped persecuted mathematicians. A few days after the Arrow-Cross takeover, on October 19, 1944, Dénes Kőnig committed suicide.

Adolf Szücs studied in Budapest and Paris, taught in high school, and in the late 1920s he received an appointment at Gusztáv Rados's department at the Budapest Technical University. His research concerned variation calculations and differential equations. On February 3, 1945, a group of the Arrow-Cross took him from his home. He was last seen on February 4, 1945.

Pál Csillag studied with Lipót Fejér and earned his doctorate when he was 21 years old. The Budapest Goldberger Textile Works employed him as a mathematician, but they had to retire him in 1938 because he was Jewish. He perished in 1944, but there is no exact information about his death.

Géza Grünwald (1910–1942)[4] studied in Szeged and earned his doctorate in 1935. Approximation theory was his principal research interest. He was member of a circle of young mathematicians that held weekly meetings in the City Park at the statue of Anonymus. He was murdered in one of the slave labor camps that were often indeed death camps. Today, the Bólyai János Mathematical Society has a Géza Grünwald memorial medal. This annual award is given to mathematicians under 30 years of age who have demonstrated considerable achievements in fundamental mathematical research.

Géza Grünwald.
Courtesy of Éva Gergő, Budapest.

Dezső Lázár.
From Középiskolai Matematikai Lapok. <http://www.komal.hu/tablok/?fenykep=150>

Dezső Lázár[5] began his studies in Budapest, but had to continue in Szeged because of the *numerus clausus* law. He could not find employment following graduation and worked as an apprentice to a cabinet maker. Then, he was offered a job in Kolozsvár (today, Cluj-Napoca in Romania) which had again become part of Hungary. In 1942, he was ordered for slave labor. He had one printed publication while he lived; it was in set theory, in the early 1930s. Pál Erdős found this paper important and showed it to John von Neumann who arranged for its publication in the periodical *Compositio Mathematica*. In 1947 Lázár had another publication, posthumously, arranged for him by his surviving friends.

We learned more about Dezső Lázár from the renowned late mathematician László Fejes Tóth:[6]

My dear friend, Dezső Lázár, turned my attention to the area of research that I worked in during my entire life; this was the project of arranging and covering the surface. . . . About Dezső Lázár I would like to say that when I moved to Kolozsvár, he was there, working as a teacher in the Jewish Gimnázium. Later, he was called upon forced labor service. He was made to detect mines, was wounded in the thigh, and left to bleed to death. While he was in the forced labor service, we kept in close contact with his family. We often visited his wife and two small children. We learned about what happened with him from his wife. I don't remember when this exactly happened, because the years have become blurred in my memory. His wife was a refined, beautiful lady and the thought horrifies me to this day that she was dragged away in a box-car and after a lot of suffering they murdered her with her two small children in the gas chambers of Auschwitz.[7]

We conclude with a few words about the mathematician Dániel Arany (1863–1945) who, while working as a high school maths teacher in 1893, founded a mathematics magazine for high-school pupils, *Középiskolai Matematikai Lapok*. He edited this for a few years; then, he passed the editorship over to László Rátz. The magazine has fostered young talent in mathematics.

Between 1905 and 1919, Arany taught in a technical college, but after 1919, he was forced into retirement for alleged involvement in the communist dictatorship. He never found employment again. He continued his research in probability theory and game theory, and co-authored a monograph in actuarial mathematics. His most important contribution remained in the area of high school mathematical education.

In 1944, the Jewish Arany and his wife were incarcerated in the ghetto where they both died. Their remains rest in a mass grave of Jewish victims in the garden of the Dohány Street Synagogue. Before moving into the ghetto, Arany donated his valuable collection of mathematical books to the Eötvös Loránd Mathematical and Physical Society. Today, the mathematical competition of 1st and 2nd grade high school pupils carries Dániel Arany's name.

Mass graves of Holocaust victims, among them Dániel Arany and his wife, in the garden of the Dohány Street Synagogue, with his portrait.

Eszter Bóra, "Ismeretlen ismerősünk Arany Dániel" (Our unknown friend Dániel Arany). <http://matek. fazekas.hu/portal/kutatomunkak/Bora_Eszter/AD_ismeretlenismeros.pdf> (downloaded November 15, 2013).

Saviors

Gábor Sztehlo

Throughout the tragedy of the Hungarian Holocaust, a few individuals rose to become heroes and saved many lives. We single out two among them and

Plaque on the wall of the Lutheran church, dedicated to Gábor Sztehlo's memory, at the Bécsi kapu Square, District I.

mention some future scientists who were among those they saved. Gábor Sztehlo was one of the saviors, and George A. Olah (see Chapters 1, 6, and 7) was among his charges. Olah was in the last year of his high school studies in 1944. When he and his wife immigrated to North America, they vowed to break with the sad past, which did not mean that they would not observe keenly and critically recent developments in Hungary. They have found it especially painful that the Hungarian responsibility for past tragedies has still not been faced.[8]

Gábor Sztehlo (1909–1974) was a Lutheran minister who responded to Bishop Sándor Raffay's (1866–1947) call to save persecuted Jewish children who had converted, and Sztehlo organized protective homes for them. Soon, he extended his efforts to all Jewish children and eventually to all children that he found abandoned as he continued his activities after liberation. Only the Swiss Red Cross assisted his endeavor. In 1972, Yad Vashem granted Sztehlo the title "Righteous among the Nations."

The Sztehlo statue on Deák Square is a beautiful creation of art; it shows quite different images when viewed from different angles.

Two views of the Gabor Sztehlo statue on Deák Ferenc Square, District V, by Tamás Vigh and Barnabás Winkler (2009). Inscription: GABOR SZTEHLO (1909–1974), LUTHERAN PASTOR, SAVED WITH GOD'S HELP AROUND 2000 CHILDREN AND ADULTS DURING THE RULE OF THE FASCIST ARROW CROSS PARTY, AND LATER GAVE THE ORPHANS HOME, FAITH, AND DIGNITY.

Raoul Wallenberg

Raoul Wallenberg is the best-known savior of Holocaust victims in Budapest. Among the thousands that he and his helpers saved, there were noted future scientists. Wallenberg came to Budapest in July 1944 and joined the Swedish Embassy as a low-ranking diplomat. By then, Hungarian Jewry had been largely deported and exterminated, but there were still about 200,000 Jews in Budapest. Wallenberg was 32 years old; he was a member of a famous banking family. In Sweden, he worked for a commercial company that had Hungarian contacts. The head of the company suggested to the Swedish ministry of foreign affairs that Wallenberg should be sent to Hungary because they knew about his interest in humanitarian matters. Sweden's behavior during the war was not exemplary and there were voices in Sweden that the country should do something to balance its political actions. Wallenberg's heroism went far beyond any definition of duty.

During the last days of the battle of Budapest, Wallenberg, his secretary, and his driver lived in the cellar of a bank. There was also the family of Lars (László) Ernster, who we mention later.[9] Ernster's father-in-law was a manager of the Orion radio company, which had a good business relationship with Sweden. The

Left: Statue of Raoul Wallenberg, St. István Park, District XIII. *Right*: The highlighted detail shows Wallenberg's portrait and the text says: Raoul Wallenberg Envoy of the Swedish Nation. The original statue was by Pál Pátzay (1949) and the copy standing at St. István Park is by Sándor Győrfi (1999).

RAOUL WALLENBERG

SVÉD DIPLOMATÁT
ÜLDÖZÖTT MAGYAR ZSIDÓK EZREINEK
MEGMENTŐJÉT 1945 JANUÁRJÁBAN EBBŐL
A HÁZBÓL HURCOLTÁK EL A SZOVJETUNIÓBA
AHOL A SZTÁLINI TERROR ÁLDOZATA LETT.

DER SCHWEDISCHE DIPLOMAT, DER RETTER VON
TAUSENDEN VERFOLGTEN UNGARISCHEN JUDEN
WURDE IM JANUAR 1945 AUS DIESEM HAUS IN DIE
SOWJETUNION VERSCHLEPPT WO ER DEM TERROR
DES STALINISMUS ZUM OPFER FIEL

RAOUL WALLENBERG EGYESÜLET

Plaque on the house from
which Raoul Wallenberg
left to meet the Soviets,
who abducted him in
January 1945, 16 Benczúr
Street, District VI.

family was under Wallenberg's protection and Ernster's father-in-law was one of Wallenberg's associates in his activities. On January 17, 1945, when the Soviet troops were only a few streets away from their hiding place, Wallenberg decided to go over to the Soviets to inform them about what had happened in Budapest. Nobody ever saw him again. The Soviets arrested him, and he perished somewhere in the Soviet Union.

After World War II, there was a movement to commemorate the hero Wallenberg in Budapest. Annie Fischer, the world-renowned pianist, who survived the Holocaust in Sweden, initiated the movement donating the proceeds of her first concert upon her return to Budapest straight after the war. A statue was commissioned, which Pál Pátzay created, and the date for unveiling was set for April 9, 1949. In the meantime, Wallenberg's name became a political liability in Soviet-dominated Hungary as his disappearance pointed to criminal action by the Soviet Union. Only hours before the appointed ceremony, the statue was brutally removed, and it vanished, only to resurface a few years later in the town Debrecen. There, it stands in front of the BIOGAL pharmaceutical plant. The statute, showing a young man struggling with a snake, symbolizes the righteous overcoming evil. In an alternative interpretation, it is the right medicine overcoming illness. For decades, the snake slayer remained identified with Wallenberg's name.

With the gradual loosening of the pro-Soviet communist political system in the late 1980s, increasingly there was talk of bringing back the Wallenberg statue to Budapest to its original location in St. István Park. However, in addition to the Debrecen public having grown close to this sculpture, the figure had become a symbol of BIOGAL products. A compromise was reached to prepare a copy of the original statue, furnish it with Wallenberg's portrait and proper inscriptions, and to erect it in St. István Park. This statue was unveiled on April 18, 1999, 50 years after the original date.

Statue dedicated to Raoul Wallenberg at 80 Üllői Avenue, District VIII.

Last known photograph of Raoul Wallenberg.

Courtesy of the late Lars Ernster.

Left: Plaque, by Antal Czinder, commemorates the office of Wallenberg and his associates at the end of 1944, at 6 Harmincad Street, District V. Today the building is the British Embassy.

Below: Street sign and relief (by Gergő Gábor Bottos) honoring Raoul Wallenberg, at 11 Raoul Wallenberg Street, District XIII.

EBBEN AZ ÉPÜLETBEN
TALÁLTAK REJTEKHELYET
RAOUL WALLENBERG
ÉS MUNKATÁRSAI 1944·VÉGÉN
•
AT THE END OF 1944
IT WAS IN THIS BUILDING
THAT RAOUL WALLENBERG
AND HIS COLLEAGUES
FOUND REFUGE TO CARRY OUT
THEIR LIFESAVING WORK

SPONSORED BY THE
WALLENBERG EGYESÜLET
FROM PUBLIC CONTRIBUTIONS

In the meantime, another variant of the statue showing the young man overcoming a snake had been put in front of a university medical school clinic, the Department of Diagnostic Radiology and Oncotherapy. Pátzay prepared it as a half-size version of the original. When it was unveiled, again, it was to symbolize healing rather than Wallenberg. Eventually, however, it was augmented with a tablet showing that it represented Raoul Wallenberg.

In 1987, a magnificent new memorial was erected on the spacious Szilágyi Erzsébet Alley, District II. Its story was also not without complications. The statue had already been completed when the Hungarian authorities were still wary about the political correctness of having such a monument in such a busy location. They were still afraid of a Soviet reaction against honoring a hero who had fallen victim to Soviet crimes and the strengthening overt anti-Semitism may have worried them as well. For a while, during this period of hesitation, the monument found temporary home in the garden of the residence of the United States Ambassador.

Raoul Wallenberg memorial on Szilágyi Erzsébet Alley, District II, by Imre Varga, 1987.

The front of this memorial shows Wallenberg in between two pieces of stone as though he had split them. On the back of the two large stones between which Wallenberg stands, there is a carving in gold: it shows the same young man overcoming the snake, just as on the other statues.

Bob Weintraub, librarian of an Israeli college, compiled the stories of five future chemists who, having been saved by Wallenberg, then achieved notable careers in science.[10] We have added a couple of other scientists to Weintraub's

Left: Lars Ernster in 1978, on the left, behind the Queen of Sweden. *Right*: John Paul II receiving Ernster at the Vatican.

Courtesy of the late Lars Ernster.

Miklós Bodánszky in 1999 in Princeton.

Photo by and courtesy of Eszter Hargittai, Evanston, IL.

Margaret Demény in 2007 in New York City.

Gabor Somorjai in 2000 in Budapest.

Francis Dov Körösy in 1963 at the Negev Institutes for Desert Research.

David Tuviyahu Archives of the Negev, Ben-Gurion University of the Negev. Courtesy of Bob Weintraub, Beer Sheva, Israel.

Andrew F. Nagy.

Courtesy of Andrew F. Nagy, Ann Arbor, MI.

Janos Hajdu.

Courtesy of Janos Hajdu, Cologne.

list. Lars Ernster (1920–1998) has already figured earlier. In Hungary, he had been excluded from higher education. In 1946, he and his violinist wife left for Sweden. He studied at the University of Stockholm and earned his PhD degree. Biochemistry and cell biology became his principal research areas. He was a professor of biochemistry at the University of Stockholm and a member of the Royal Swedish Academy of Sciences. He served on the Nobel Committee for Chemistry between 1977 and 1988, and he was a member of the Board of Trustees of the Nobel Foundation in 1990 and 1991. Mrs Ernster, Edit, was for a long time, the first violinist of the Stockholm Opera.

Miklós Bodánszky (1915–2007) graduated from the Budapest Technical University in 1939 but was excluded from further studies. He survived unemployment and the slave labor service. He spent the last portion of the period in hiding, and Wallenberg saved him. He earned his doctorate in 1949, left Hungary in 1956, and had a bright career in organic chemistry at the Case Western Reserve University in Cleveland, OH. His monographs on peptide chemistry have been used worldwide.

Margaret Demény studied chemistry. In 1942, she was able to get her doctorate thanks to a supportive professor at the University of Budapest, Gyula Gróh (Chapter 5). The Nazis murdered her husband, but she survived under Wallenberg's protection. In 1947, she moved to the United States, and devoted her professional life to biochemical research on spinal cord injuries at the Mt. Sinai Hospital in New York City and at New York University.

Wallenberg saved a nine-year old boy, Gabor Somorjai (1935–). In 1956, he immigrated to the United States and became a world leader of surface science and catalysis. He has spent much of his professional life at the University of California at Berkeley and the Lawrence Berkeley National Laboratory. Among his highest distinctions, he has received the Wolf Prize in Chemistry and the US National Medal of Science.

Francis Dov Körösy (Ferenc Kőrösy, 1906–1997) earned his PhD in chemistry in 1928, in Budapest. He worked for Tungsram. During the Hungarian Holocaust, he possessed Swedish papers, but these would not have saved him from being shot into the Danube. At the last minute, however, Wallenberg arrived on the scene and scared away the Arrow-Cross people. In 1957, Körösy moved to Israel where he served as director of the Chemical Laboratory at the Negev Institutes for Desert Research.

Andrew F. Nagy's (1932–) father was killed in Auschwitz. Andrew and his mother survived the last months in Budapest in a house protected by Wallenberg. In 1949, he left Hungary. First he studied in Australia, then in the United States. He has been at the University of Michigan at Ann Arbor, and is now Professor Emeritus of the Department of Atmospheric, Oceanic and Space Sciences.

Janos Hajdu's (1934–) father perished in a slave labor camp. Janos was protected in Budapest by Wallenberg. He studied physics in Budapest and Göttingen, and lives in Germany. He is a physicist, Professor Emeritus of Cologne

University; he is internationally renowned for his research on various aspects and applications of quantum theory. In 2004, he was elected as an external member of HAS.

Notes

1 Lajos Pillich, "Richter Gedeon (1872–1944)." In Krisztina Novák Takács and István Hermecz, Eds., *Esti beszélgetés: Magyar gyógyszerészkutatók portréi* (Evening Conversations: Portraits of Hungarian Pharmacologists, Budapest: Hungarian Society of Pharmacology, 2005), pp. 69–82.

2 Max Born, *My Life: Recollections of a Nobel Laureate* (New York: Charles Scribner's Sons, 1978), p. 214.

3 Ferencné Nyitrai, "Fellner Frigyes (1871–1945)" ("Professional activity of Frigyes Fellner"). In *Nagy magyar statisztikusok 18: Fellner Frigyes (1871-1945) műveinek válogatott bibliográfiája* (Budapest: KSH Könyvtár és Dokumentációs Szolgálat, 2001), pp. 7–20.

4 These are the correct years of birth and death communicated by Géza Grünwald's daughter, Éva Gergő.

5 László Filep, "Magyar matematika Erdélyben a két világháború között." *Magyar Tudomány* 2001/5.

6 István Hargittai, "Fejes Tóth László." *Magyar Tudomány* 2005, 166(3), 318–324, actual quote, p. 319.

7 IH adds: In 1942, my father was killed the same way as Dezső Lázár. Then, our story continued as his family's, but only up to a point. In June 1944, my mother, my brother, and I were put into a box-car and the train started for Auschwitz. Along the way, somewhere, however, the train stopped, moved back for a while, and continued in a different direction, toward Vienna. What happened was that a train that had been destined for Austria had already left for Auschwitz, by mistake. Our train was the replacement. My brother was ten years old and I was not yet three [I. Hargittai, *Our Lives: Encounters of a Scientist*, Budapest: Akadémiai Kiadó, 2004; pp. 52–54].

8 E-mail message from George A. Olah, September 10, 2013.

9 István Hargittai, *Candid Science II: Conversations with Famous Biomedical Scientists* (edited by Magdolna Hargittai, London: Imperial College Press, 2002), "Lars Ernster," pp. 376–395.

10 Bob Weintraub, "Five Chemists Whose Lives Were Saved by Raoul Wallenberg." *Bulletin of the Israel Chemical Society* 2009, December, pp. 52–57.

APPENDIX: DICTIONARY

English	Hungarian	Hungarian	English
alley	fasor	fasor	alley
avenue	út	folyó	river
boulevard	körút	kert	garden
cemetery	temető	kerület	district
circus	körönd	körönd	circus
district	kerület	körút	boulevard
embankment	rakpart	köz	passage
garden	kert, liget	liget	gardens, park
gardens	liget	park	park
island	sziget	rakpart	embankment, quay
necropolis	sírkert	sétány	walkway
park	liget, park	sírkert	necropolis
passage	köz	sziget	island
quay	rakpart	temető	cemetery
river	folyó	tér	square
square	tér	út	avenue
street	utca	utca	street
walkway	sétány		

INDEX

Index

Index

composers 16, 49, 51
Compositio Mathematica 286
Compromise (1867) 31, 37, 60, 61, 104, 108, 111, 113, 169
computers 182, 197, 208, 246, 259, 266
concentration camps 75, 161, 197, 273, 279, 282–284
conductive education. *See* Pető Institute
Congressional Human Rights Caucus (US House of Representatives) 259
continuing education 118, 188
conversion of religion 45, 232, 233
cooling techniques 199
Copenhagen 88, 250
Copernicus 7
Corvinus University 147, 149–151
Crafoord Prize 17
critical mass, 254
Croatia 73
crystallographers 130, 192
Csáki, Frigyes 203, 281
Csalán Avenue 50
Csángó 46, 47
Cserháti, Sándor 155
Csillag, Pál 284, 285
Csillebérc 202, 203
Csömöri Street 263
Csonka János Museum. *See* museums
Csonka, János 190, 212, 215, 216
Csonka, Pál 190, 191
Csorba 110
Csűrös, Zoltán 196, 197
cultural superiority, policy of 74
Curie, Irène 203
current generator 183
Czakó, Adolf 2, 3

D

Dalton, John 65, 142
Damjanich Street Gimnázium. *See* gimnáziums
Damjanich Street 121, 185, 257
Danner, János 205
Danube River
Arrow-Cross murders at 269, 270, 274, 283, 294
bridges of 177, 214, 249

designing embankment of 173
drowning in 258
regulation of 172
Shoes memorial 269
Danube Iron Works 190
Darjeeling 40
Darkness at Noon (Koestler) 18, 21
Dartmouth College 259
Darwin, Charles 64
Davy, Humphry 66
Deák Ferenc Square 288
Deák, Ferenc 30, 31
Debrecen 14, 47, 120, 188, 290
deficiency diseases 272
Délibáb Street 49, 92
Dembinszky Street 138
Demény, Margaret 293, 294
democratic periods 77, 256
Denmark 12, 87, 88
deportations 14, 16, 75, 197, 270, 273, 275, 276, 283, 289
Déri Miksa Street 183
Déri, Miksa 183
Descartes, René 26, 27
Detre, László 86
differential equations 285
Diószegi, Sámuel 169
Dirac, Paul 84
disinfection 99
Dísz Square 132
Divald, Kornél 25, 55
Djerassi, Carl 79
Dob Street 78, 79, 223
Dohány Street Synagogue 270, 286, 287
Dollinger, Gyula 116
Domokos, Pál Péter 46
Donáth, Ferenc 157
Dorottya Street 39
Dózsa György Avenue 49
Dracula 163
Duna Street 275
dynamo 59

E

ear research. *See* otology
earthquake research 84, 85
ecologists 164, 237
economists 47, 149, 154, 160, 163, 232, 250, 278, 281, 282
economics 14, 31, 154, 200, 223

Ecséd 271
Edison, Thomas A. 224
Edömér Street 199
Egerváry, Jenő 190
Egressy, Béni 39
Egyed, László 85
Egyetem Square 62, 69
Egypt 40, 251
Einstein, Albert 10, 68, 90, 238, 240, 254, 258
electrical engineers 183, 186, 197, 199, 202, 282
electricity 59, 209, 217
electrification 186, 217
electrocardiography 128
electronics 197
emancipations 47, 62, 95
emigration 94, 160, 232
energy studies 199
Engels, Friedrich 112
engineer-physicists 232
England. *See* Great Britain
Enlightenment 78, 230
Entz, Ferenc 146, 147
Eötvös Avenue 245
Eötvös József Square 31, 60
Eötvös Loránd Geophysical Institute 69
Eötvös Loránd Mathematical and Physical Society 239, 284, 286
Eötvös Loránd University 17, 57–96, 167, 202, 246
Eötvös, József 31, 60, 64, 97
Eötvös, Loránd 17, 31, 57, 60, 67–69, 76, 84, 87, 90, 274, 277
epidemiology 107
epistemology 250
Erdei, Ferenc 28, 152
Erdey, László 200, 201
Erdey-Grúz, Tibor 82
Erdős Number 94
Erdős, Ervin G. 281, 282
Erdős (Erdos), Pál (Paul) 91–93, 286
Ernst, Jenő 281
Ernster, Edit 294
Ernster, Lars (László) 289, 290, 293–295
Erzsébet Boulevard 114
Erzsébet Bridge 261
Eszék Street 46
Esztergom 57
Eszterházy Street 64, 175
ETH. *See* Swiss Federal Institute of Technology

302

Index

Index

Index

Olah, George A. 14, 15, 180, 181, 232, 261–263, 268, 288
Olivecrona, Herbert 134
ophthalmologists 105, 106, 115, 123, 129, 132, 133
ophthalmology 123, 132, 133
organotherapia 272
Orion radio company 289
Ormos, Imre 148
ornithologists 164
Országh, László 47
Orszag-Land, Thomas 281, 283
orthopedics 116
Ortvay, Rudolf 83, 84, 90, 280
Orvosi Hetilap (Medical Weekly) 104
otology 117
Ottoman Empire. *See* Turkey

P

Pach, Zsigmond Pál 281
painters 45, 112, 209, 257, 258
Palotás, László 199
Pándi (Kardos), Pál 282
Papnövelde Street 230
Paris world exhibition 216
Paris 14, 138, 160, 216, 224, 285
Parliament building 32, 33, 59, 152, 153, 249
Parliament 24, 34, 74, 122, 173
pathologists 109, 118, 127, 129, 144
pathology 98, 105, 109, 123, 128, 130, 145
Pattantyús-Ábrahám, Géza 188
Paul Street Boys (Molnár) 167
Pauler Street 44, 193
Pauncz, Ruben 282
Pázmány Péter Catholic University 76
Pázmány Péter Walkway 17
Pázmány, Péter 57, 58, 76
Péch, Antal 220
Péch, Géza 205
Pécs 117
Pecz, Samu 71, 72
pedagogues 51, 133, 134, 188, 197, 207, 230, 235
pediatricians 109–111, 119
pediatrics 109–112, 119
period between the two world wars. *See* Horthy era

Persia 40, 42
Personal Knowledge (Polanyi) 250
Pesti Barnabás Street 261
Pesty, Frigyes 44
Pethe, Ferenc 152, 163
Pető Institute 133, 134
Pető, András 133, 134
Petőfi Square 45
Pfeifer, Ignác 184, 185, 276
pharmaceutical chemistry 186
pharmaceutical companies 271–274, 290
pharmaceuticals 81, 82, 186
pharmacies 271, 272
pharmacists 14, 82, 127, 148, 232, 271, 272
pharmacologists 127, 144, 146, 281, 282, 295
pharmacology 105, 111, 127, 295
philology 29, 43
philosophers 26, 27, 65, 78, 149, 250, 256, 281
philosophy 29, 31, 60, 61, 71, 73, 75, 76, 78, 91, 250
photocopying 278
physical chemists 49, 82, 192, 250
physical education 131
physical geography 29
physicians. *See* medical doctors
physicist-engineers. *See* engineer-physicists
physicists 74, 267, 276
  Békésy, Georg von 12, 13, 86, 192–195
  Bethe, Hans 242
  Born, Max 74, 96, 276, 295
  Bródy, Imre 276, 277
  Eötvös, Loránd 17, 31, 57, 60, 67–69, 76, 84, 87, 90, 274, 277
  Gabor, Dennis 13, 195, 232, 259, 260, 276
  Gay-Lussac, Louis 66
  Gribov, Vladimir N. 94–96
  Gyulai, Zoltán 188
  Hajdu, Janos 294
  Helmholtz, Hermann von 67, 142
  Jedlik, Ányos 17, 59, 212
  Joliot-Curie, Frédéric 203
  Kövesligethy, Radó 84, 85, 280
  Kurti, Nicholas 251
  Lánczos, Cornelius 90, 91
  Marx, George 235, 241

Ortvay, Rudolf 83, 84, 90, 280
Selényi, Pál 277, 281
Simonyi, Károly 202, 228
Szilard, Leo 17, 88, 232–234, 238, 240, 252–255, 260, 264, 268
Tangl, Károly 83, 84, 188, 192, 280
Teller, Edward 17, 18, 75, 180–182, 232–234, 246, 248–250
Wigner, Eugene P. 13, 39, 56, 75, 84, 180, 181, 229, 232, 234, 236–242, 246
pianists 49, 290
Piarist Order 67, 76, 87, 120, 261–263
Piarista Gimnázium (Gimnázium of the Piarist Order). *See* gimnáziums
pilgrimages 40, 45
Planet Mars. *See* Mars
plant sciences 137, 148, 154, 162, 164, 272
poets (poetry) 29, 36, 38, 39, 134, 256, 283
Pogány, Béla 182, 188, 280
Pogány, Frigyes 200
Poland 42
Polanyi family 251
Polanyi, Adolf 250
Polanyi, John C. 14, 242
Polanyi, Karl 250
Polanyi, Laura 250
Polanyi, Magda 14
Polanyi, Michael 14, 238, 242, 250, 276
Polgár, Judit 259
Polgár, László 259
Polgár, Sofia 259
Polgár, Susan 259
Politechnicum 173
politicians 31, 37, 60, 61, 78, 82, 149, 150, 151, 160, 189, 203, 211
Polya, George 74, 259
polyglots 40
polymaths 27, 137
postage stamps 11, 12, 207, 227
Pozsony (now, Bratislava, Slovakia) 10, 24
Pozsonyi Avenue 54, 55, 273
Pressburg. *See* Pozsony
Priestley, Joseph 65
Princeton 242, 245, 246, 293

310

Index

Index

INDEX OF ARTISTS
AND ARCHITECTS

Index of Artists and Architects

ALSO BY THE AUTHORS

Magdolna Hargittai, *Women Scientists: Reflections, Challenges, and Breaking Boundaries* (Oxford University Press, 2015)

Balazs Hargittai, M. Hargittai, and Istvan Hargittai, *Great Minds: Reflections of 111 Top Scientists* (Oxford University Press, 2014)

I. Hargittai, *Buried Glory: Portraits of Soviet Scientists* (Oxford University Press, 2013)

I. Hargittai, *Drive and Curiosity: What Fuels the Passion for Science* (Prometheus, 2011)

I. Hargittai, *Judging Edward Teller: A Closer Look at One of the Most Influential Scientists of the Twentieth Century* (Prometheus, 2010)

M. Hargittai and I. Hargittai, *Symmetry through the Eyes of a Chemist*, 3rd Edition (Springer, 2009; 2010)

M. Hargittai and I. Hargittai, *Visual Symmetry* (World Scientific, 2009)

I. Hargittai, *The DNA Doctor: Candid Conversations with James D. Watson* (World Scientific, 2007)

I. Hargittai, *The Martians of Science: Five Physicists Who Changed the Twentieth Century* (Oxford University Press, 2006; 2008)

I. Hargittai, *Our Lives: Encounters of a Scientist* (Akadémiai Kiadó, 2004)

I. Hargittai, *The Road to Stockholm: Nobel Prizes, Science, and Scientists* (Oxford University Press, 2002; 2003)

B. Hargittai, I. Hargittai, and M. Hargittai, *Candid Science I–VI: Conversations with Famous Scientists* (Imperial College Press, 2000–2006)

I. Hargittai and M. Hargittai, *In Our Own Image: Personal Symmetry in Discovery* (Plenum/Kluwer, 2000; Springer, 2012)

I. Hargittai and M. Hargittai, *Symmetry: A Unifying Concept* (Shelter Publications, 1994)

Ronald J. Gillespie and I. Hargittai, *The VSEPR Model of Molecular Geometry* (Allyn & Bacon, 1991; Dover Publications, 2012)